A potted history of
Vegetables

LORRAINE HARRISON

Ivy Press

First published in Great Britain 2011 by

Ivy Press

210 High Street, Lewes,

East Sussex, BN7 2NS, UK

www.ivypress.co.uk

British Library Cataloguing-in-Publication Data
A CIP catalogue record for this book is available from
the British Library.

ISBN 978-1-907332-61-6

Ivy Press

This book was conceived, designed and
produced by Ivy Press
Creative Director *Peter Bridgewater*
Publisher *Jason Hook*
Editorial Director *Caroline Earle*
Art Director *Michael Whitehead*
Senior Editor *Stephanie Evans*
Designer *Richard Constable*

Printed in China
10 9 8 7 6 5 4 3 2 1

Colour origination by Ivy Press Reprographics

A potted history of
Vegetables

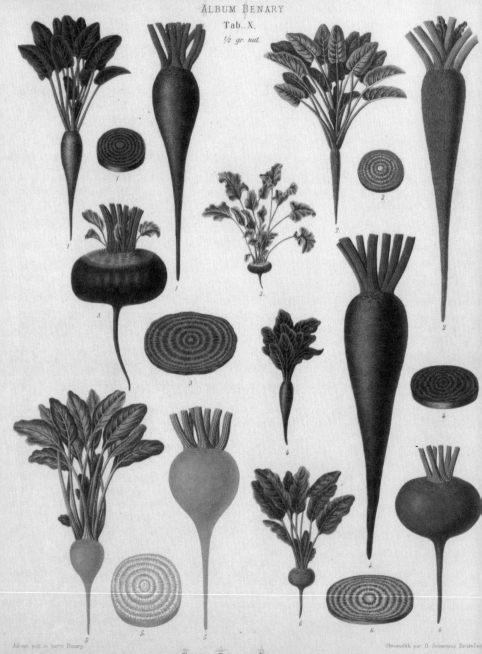

Ad nat. pict. in horto Benary.

Chromolith par G. Severeyns, Bruxelles.

ERNST BENARY, ERFURT.

"All our progress is an unfolding, like the vegetable bud,
you have first an instinct, then an opinion, then a knowledge,
as the plant has root, bud, and fruit. Trust the instinct to
the end, though you can render no reason."
RALPH WALDO EMERSON

FOREWORD

S I BEGAN TO READ THIS ENGAGING AND REWARDING little book I was transported back to August 2009, to a week sweltering in New York City. I was staying at the famous (infamous?) Hotel Chelsea on West 23rd Street, a sprawling ten-storey former apartment building that counts Bob Dylan, Allen Ginsberg, Jimi Hendrix, William Burroughs and Dylan Thomas among the pop and poetic literati to have sought its refuge. Mine was one of its basic rooms, sans air-con – its claustrophobic atmosphere further augmented by the presence of a giant refrigerator whose rattling motor upped the temperature while simultaneously contributing its own idiosyncratic soundtrack. There is no restaurant at the Chelsea, but it is located in a part of town peppered with outrageously bountiful delis and take-aways stuffed with garden and small-holder produce. That ancient, unsteady Kelvinator and I soon became good friends.

Today's better sort of New York deli is a Garden of Eden, and queuing to pay for exotic titbits my thoughts flew some 3,000 miles eastwards, to the English–Welsh border and my own fecund but far less colourful veg patch where peas, asparagus, broad beans, waxy potatoes and young salad leaves help to sustain us for much of the year.

My father grew lush fruit and veg for a hungry family in our Cotswold garden during the privations following the Second World War, and ever since I have taken a keen interest in the history, provenance, cultivation and eating of home-grown food – even when working in London, where my 'garden' was a single north-facing window box – growing, I recall, some excellent French tarragon. Alas, there was no Lorraine Harrison to guide me in those days, but gardeners finding themselves similarly lusting after fresh vegetables will glean much from these pages.

An ounce of love may, indeed, be worth a pound of knowledge (to paraphrase the English poet Robert Southey, writing in 1820), but by growing our own food and having some inkling of its origin kindles a concern for what we consume and demonstrates our care for the land that supports the plants we eat, adding also to our general well-being and a wider understanding of Nature itself. Put another way, don't we all the more enjoy the work of our favourite writers, singers and actors when we know something of their lives and past achievements? Yes. In some respects, therefore, this book is a fan's newsletter: a place for winkling out delightful snippets of arcane information that will, I promise, enhance your eating experiences and – more beneficially, perhaps – lead to a discovery of 'lost' heirloom varieties where tastes and textures belong to a superior, though not inaccessible, gastronomic world.

DAVID WHEELER
Editor, *Hortus*

THE PEPPER

Capsicum spp.

SOLANACEAE

WITH ITS BRIGHTLY coloured shiny fruits and sometimes hot spicy taste, it is hardly surprising that the pepper plant has held such a fascination for humankind through the centuries. First harvested in the wild around 10,000 years ago in South and Central America, it is thought the pepper was in domestic cultivation by at least 3300 BCE. The pepper arrived in Europe in the 16th century, brought back by the returning Christopher Columbus. This curious and attractive exotic plant then spread rapidly throughout the rest of the world and has become a vital ingredient in many cuisines.

Sweet peppers, which include the popular bell peppers, come in a wide array of shapes, sizes and colours including brown, green, orange, purple, red, white and yellow. As the fruits mature, their sugar content increases, making them sweeter and much more palatable (immature green peppers in particular can be bitter). Hot chilli peppers are usually smaller than the sweet type and, again, vary in shape, size and colour but, most importantly, it is the relative level of heat content that differs most between the varieties. (For instance, a habañero pepper is one hundred times hotter than a jalapeño!) Indeed, the chilli pepper has its own dedicated system of measuring the hotness for each type, known as the Scoville Scale (see page 18). As a rule of thumb, the smaller the pepper, the hotter its taste, while dried peppers are considerably more potent than their fresh counterparts.

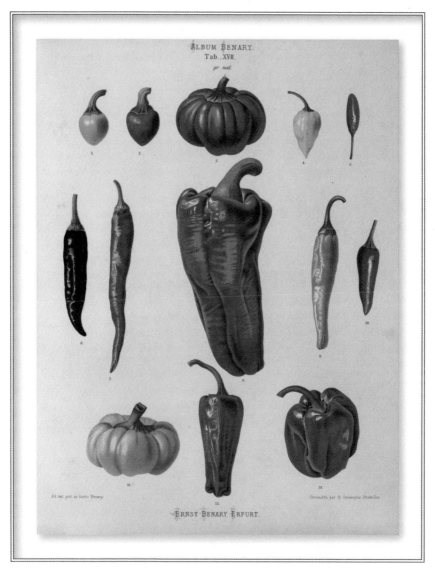

'Spanish Pepper', by Ernst Benary (1819–1893)

from Album Benary

COLOUR LITHOGRAPH BY G. SEVEREYNS

PEAS & WARTS

Legend has it that an efficacious treatment for warts is to wrap a pea in paper (one per wart) and bury it while reciting this old saying, 'As this pea shall rot away, so my warts shall soon decay.' Then, as the peas fade to nothing, so will the warts!

SEED SAVERS

When choosing seed varieties, gardeners often prize the novel and experimental over the tried and tested. Why always grow the long, green type of courgette when round and yellow ones are also available? Why restrict your choice of tomato to the round, red kind, when purple, orange, yellow, striped and heart-shaped ones taste just as good? Home growers have this choice, while shoppers rarely do. However, in recent years gardeners have expressed increasing concern over the control of the supply of seeds by major seed companies, especially of vegetables and fruit. Commercially available seeds are F1 hybrids, which means that once a plant has flowered and set seed, the seeds cannot be collected, stored and sown, because they are not viable (being either sterile or degenerate), unlike traditional open-pollinated varieties; thus new seeds have to be purchased each season.

What is an expensive irritation for gardeners in Western society is a real privation for subsistence farmers in developing regions of the world. The list of available seeds has been drastically reduced as major suppliers concentrate on developing limited plant cultivars that are best suited to large-scale commercial food production. Qualities such as taste, variety and suitability to locale all play second fiddle to the need for disease resistance, bumper crops, long-term storing properties, and the ability to be transported great distances and arrive unblemished (although often under ripe).

Thankfully, gardeners have risen up in revolt against this trend and demand for open-pollinated heritage and heirloom cultivars has resulted in a whole host of vegetable and fruit varieties being increasingly available. Many gardeners have also rediscovered the traditional art of seed saving and freely swap their harvest at local horticultural shows.

BEAUTIFUL BEANS

Few kitchen gardens look complete in summer without an arch or tall support smothered with runner beans (*Phaseolus coccineus*). Contemplating the plant's lovely and prolific red flowers, it is easy to see why it was grown purely as an ornamental plant in 17th-century England. There is some dispute as to how exactly it arrived in the country in the first place. One version has it that it was King Charles I's gardener, John Tradescant the Younger (1608–62), who introduced the runner bean from North America in 1633. An alternative source tells us that it was introduced from Mexico by Henry Compton (1632–1713), Bishop of London, in the mid-17th century. Whatever the truth, it was not until well over a century later that its culinary qualities were fully appreciated. The cultivar 'Painted Lady' is widely grown today, valued almost as much for its bicolour white and red flowers as its fine-tasting beans. Legend has it that it was named after Queen Elizabeth I, whose face was heavily caked with white chalk and rouge.

A BRIEF HISTORY OF CARDOONS

An edible thistle, the curious cardoon has been cultivated in Mediterranean regions for thousands of years, while Pliny (23–79 CE) is known to have hailed its medicinal properties. It was an old favourite in Victorian Britain, and is now grown for its striking silvery foliage.

The cardoon (*Cynara cardunculus*) and the globe artichoke (*C. scolymus*) are often mistaken for one and the same plant, although very different parts of each are eaten. It is the lower stems or ribs of the giant cardoon plant, instead of the flower head, that are consumed. First they must be blanched to remove any bitterness. To do this you need to wait until the end of the growing season then cut down the plant to about 15cm/6in above ground.

The desirable part is the new pale green growth that emerges. Once it is over 60cm/2ft high, remove any damaged growth and tie into cylindrical bunches, using soft twine. Wrap with cardboard and secure with string. (Traditionally straw and wicker were used and manure heaped outside to hasten the blanching process.) After about a month the stalks will be ready and should be cut down to just above ground level.

Only the blanched stems are prepared for cooking, not the leaves, and their taste is very similar to that of the globe artichoke. Heirloom cultivars to look out for include 'Tunisian Spiny,' which has very spiky ribs, and 'Inerme Blanc,' which is free from spines.

THE TURNIP

Brassica rapa var. *rapifera*

BRASSICACEAE

A ROOT VEGETABLE native to northern Europe, historically the turnip has not only been thought of as food for the poor, but also suffered the indignity of overcooking, resulting in its still somewhat underrated reputation. This is a shame because, if harvested in the spring when small and either grated raw in salads, sautéed quickly in butter, or lightly steamed, it is a delicious and subtle vegetable with tastes ranging from sweet to spicy.

Growers of heirloom turnips have a wide spectrum of colours to choose from, with white or pale yellow flesh encased in skins that range from green, white, or yellow to purple and black. The 19th-century British cultivar 'Aberdeen Green Top Yellow' (also known as 'Green Top Scotch') is yellow skinned at its base changing to purple, with dark green leaves. The French cultivar 'De Croissy', which dates from the same period, is pure white in skin and flesh with a fruity, spicy taste. Also with white skin and flesh is the very old European cultivar 'Early White Milan', sometimes called 'White Strap-Leaved American Stone'. Not surprisingly, its purple-skin counterpart is called 'Purple Top Milan' (or 'Red-Top Strap-Leaved American Stone'). 'Long Black', another cultivar from 19th-century France, has black skin and white flesh with a buttery and sweet taste, while Japan's 'Scarlet Ball' has red skin with attractive, red-veined foliage.

'Turnip remarkable for its size …', gouache by unknown Flemish artist (n.d.)
from Collection du Règne Végétal
BARON JOSEPH VAN HUERNE (1790–1820)

CROP ROTATION

The best home grower will endeavour to practise good crop rotation. Each year the bed, or beds, used for one type of plant should be rotated and another group of plants cultivated in that space. This should be done as part of a three-, four- or five-year cycle, depending on the scale of the plot, to prevent the soil from becoming stripped of the nutrients needed by a particular group of plants. Here is an example of a simple three-year rotation.

YEAR 1
Bed 1 – *Group A:* beans, celery, sweetcorn, onions, leeks, lettuce, peas, spinach, tomatoes, courgettes.
Bed 2 – *Group B:* Brussels sprouts, broccoli, cabbages, cauliflowers, kohlrabi, radishes, swede, turnips.
Bed 3 – *Group C:* beetroot, chicory, carrots, Jerusalem artichokes, parsnips, potatoes.

YEAR 2
Bed 1 – *Group C*
Bed 2 – *Group A*
Bed 3 – *Group B*

YEAR 3
Bed 1 – *Group B*
Bed 2 – *Group C*
Bed 3 – *Group A*

SOME POTATO FACTS

In an effort to make the potato accepted as food fit for a queen, rather than as mere animal fodder, Antoine-Auguste Parmentier (1731–1813) presented a bouquet of potato flowers to Marie-Antoinette. The French government had forbidden its cultivation in 1748, thinking it responsible for all kinds of health problems and it was not until 1772 that the potato was officially declared edible. Cultivars are numerous and differentiated by their seasons: earlies, second earlies, early maincrop and maincrop.

Potatoes should be stored in a cool dark place but never in a refrigerator (too cool a temperature turns their starch content to sugar). To retain their maximum nutrients (which include vitamin C, the B vitamins, potassium, calcium and iron), do not peel potatoes because most of their valuable nutrients lies just beneath the skin, which also provides fibre.

GOOD COMPANIONS

Medieval gardeners believed that a 'sympathetic magic' existed between certain plants. When grown together, the resulting crops of these empathetic partners apparently tasted better or proved more resilient to pests and predators. Modern gardeners often grow French marigolds and basil alongside their tomatoes, because both the health and taste of the resultant crop seem much improved by this companion planting method. The vibrant gold of the lovely marigold is certainly attractive to the hoverfly, an enthusiastic consumer of potato and cabbage pests. Other taste-improving combinations include carrots and onions, corn and potatoes, and chervil and radishes. Carrot growers should also sow a few rows of chives among their crops because the pungent smell of the herb deters the marauding carrot fly.

Notwithstanding the potential benefits of growing these particular combinations, it is widely acknowledged that the more species diversity that exists within a growing system, the greater the health and vigour of the plant, insect and animal life. And that has to be good!

VICTORY GARDENS

In March 1917, just prior to entering the First World War, the United States War Garden Commission was introduced. Their aim, much like that of the British 'Dig For Victory' campaign, was to encourage American citizens to grow food to help the war effort. A poster of the time proclaims 'Will you have a part in victory?' and shows a woman draped in the Stars and Stripes sowing victory-winning seeds (perhaps of the heirloom leek 'American Flag'?). The Victory Garden movement was revived in the Second World War when the equally rousing term Liberty Garden was also used. By the end of the war, the US School Garden Army had 1,500,000 child members.

More recently there has been a resurgence of the idea with the creation in 2008 of a Victory Garden at the San Francisco Civic Center. Here, ardent vegetable growers have redefined victory in the context of urban sustainability and the need to grow food 'at home for increased local food security and reducing the food miles associated with the average American meal'. Perhaps its modern-day apogee was reached the following year when First Lady Michelle Obama, aided by a team of spade-bearing schoolchildren, dug up part of the White House lawn to create a 21st-century Victory Garden. The idea was inspired by the Kitchen Gardeners International campaign 'Eat the View', which encourages the creation of edible landscapes. As all gardeners know, what goes around comes around ...

THE TOMATO

Lycopersicon lycopersicum

SOLANACEAE

IT WAS THE INCAS and Aztecs who first harvested and consumed the wild forms of tomato that grew in the valleys of the Peruvian Andes. By around 500 CE, Native Americans were cultivating and domesticating the plant. Not until the 16th century were Europeans able to sample the delights of this most delicious vegetable, when returning Spanish conquistadores introduced seeds from Mexico and Central America. The Spanish, Portuguese and Italians were the first to fully appreciate this new arrival. Other nations were much slower to try them, often distrusting the brightly coloured tomatoes and treating them as ornamental, instead of edible, vegetables.

Surprisingly impassioned battles have raged in past times over the exact definition of the tomato: Is it a fruit or a vegetable? The US Supreme Court finally settled the vexed question in 1893, ruling that it is indeed a vegetable.

The home gardener really has an advantage over commercial growers when it comes to tomatoes, because so many exciting and tasty varieties can easily be sown from seeds. Shapes range from the round, to heart, pear or plum, while colours include red, pink, orange, yellow, white, green, purple and bicoloured or striped. They may be large and heavy, such as the beefsteak types, or bite-size, such as the cherries. In addition to the numerous ways in which tomatoes can be prepared in the kitchen, their nutritional value and disease-preventive qualities are virtually unrivalled.

'Tomatoes', by Ernst Benary (1819–1893)

from Album Benary

COLOUR LITHOGRAPH BY G. SEVEREYNS

THE CHILLI HEAT SCALE

Measuring the relative heat of chilli peppers (and foods derived from them) has become something of a science. In 1912 Wilbur L. Scoville (1865–1942) devised the Scoville Organoleptic Test to measure the concentration of the chemical compound capsaicin (the thing that burns your mouth). Now known as the Scoville Scale, it is still in use, although high-pressure liquid chromatography is a more accurate way of determining heat. The relative heat of chilli peppers varies greatly, even within the same variety. Factors such as hours of sunlight, growing temperature, moisture, soil type and fertility can all play a part. Here is a sample of a few chilli varieties and their rating in Scoville heat units (SHU):

Habañero (*Capsicum chinense* 'Jacquin')
    100,000–350,000
Scotch bonnet (*C. chinense*)
    100,000–325,000
Thai pepper (*C. annuum*)
    50,000–100,000

Cayenne pepper (*C. baccatum*) &
    Tabasco pepper (*C. frutescens*)
    30,000–50,000
Jalapeño pepper (*C. annuum*)
    2,500–5,000
Cubanelle pepper (*C. annuum*) 100–1,000

GIANTS & DWARFS

A pumpkin or squash is probably the closest thing to a triffid that gardeners can grow in a kitchen garden. A pumpkin vine can allegedly grow a staggering 1m/3ft overnight and gain 14kg/30lb in a single day. 'Atlantic Giant' (often weighing 225kg/500lb) and 'Big Max' (frequently reaching 30kg/70lb) are two pumpkin varieties commonly grown by those wanting to produce record-beating fruits. At a more modest 9kg/20lb, 'Connecticut Field', also known as 'Big Tom' and 'Yankee Cow Pumpkin', is an old Native American heirloom variety dating from before 1700. At the other end of the scale is the delightful 'Jack Be Little', a miniature orange-flesh pumpkin that both cooks and keeps well.

SOME JERUSALEM ARTICHOKE FACTS

The one unfortunate fact everyone seems to know about this root vegetable is that it causes flatulence. However, it also has other, more attractive, attributes. Despite its name, it originates from North America. Sometimes referred to as 'sunchokes' (because they are the tuber of a variety of sunflower), Jerusalem artichokes (*Helianthus tuberosus*) range in colour from beige to brownish-red, have a nutty flavour, and are highly nutritious. They should be cooked well (they make a particularly good soup) to help to alleviate their malodorous side effect; indeed Native Americans traditionally roasted them in pits for up to two days.

HOW TO GROW HEIRLOOM TOMATOES

I always maintain that if you only grow one crop, make sure it is the tomato, and the more the merrier! Choose from cherry, plum and beefsteak cultivars in a range of colours from purple to yellow, pink or red- and green-striped ones.

1. Sow seed indoors in small pots or trays. This should be done approximately six to eight weeks before the last frost normally occurs in your area. A gentle heat helps germination.
2. Once the first set of true leaves (those indented with 'teeth') appears, the plants can be transplanted into 5- to 10-cm (2- to 4-in) diameter pots.
3. Let the plants continue to grow and, if they are to be cultivated in a greenhouse, transplant them into the largest pots you have. Outdoor tomatoes need to be hardened off before planting outdoors. After all danger of frost has passed, begin to place the pots outdoors for several hours each day; this acclimatizes them to the weather before you plant them in their final position.
4. Whether transplanting small plants into larger pots, or planting outdoors, they will grow more strongly if set in the soil at a lower depth than they have been previously growing. This encourages new roots to spring from the base of the stem and anchors them more securely to the soil.
5. Always provide a firm support, such as a stake, and gently tie in the plants as they grow.
6. Pinch off any side shoots that appear between the main stalk and side branches, because this encourages more tomatoes to develop.
7. Water well and regularly, and apply a liquid fertilizer weekly. Organic gardeners often favour a diluted seaweed preparation. Erratic or insufficient watering can result in the skin of the tomatoes splitting while still on the vine. Indoor tomatoes should be misted with a water spray to encourage the flowers to set fruit.
8. To harvest, gently twist the tomatoes but do not pull off. If they come away from the plant easily, they are ripe.

THE PEA

Pisum sativum

FABACEAE

PEAS ARE AMONG the oldest of cultivated vegetables. They probably originated in Asia and were grown at least as far back as 7800 BCE. Wild strains of peas were gathered and selectively improved in the Middle Eastern region. They were prized by early growers because they were easy to cultivate, kept well once dried, and were a good source of protein (they are also high in vitamin C).

The humble pea played an important role in the scientific work of the Moravian monk Gregor Johann Mendel (1822–84), whose experiments established the revolutionary science of genetics. Mendel hybridized different varieties of peas, and discovered that he was able to predict the emergence of the dominant gene by the third generation of a plant.

Shelling peas are often referred to as garden peas, or as green or English peas. Sitting in the sun shelling peas is one of life's most restful pastimes and gardeners attest to the incomparable delicacy and sweetness of taste of a freshly shelled pea. However, the pea is one vegetable that freezes particularly well. Harvested before they are fully ripe, the peas are frozen quickly and much of their taste is preserved. Both tall-growing and bush types are available. 'Alderman' is a fine cultivar from the 19th century and will quickly cover a frame, and the bush 'Dwarf Telephone' was introduced in 1888.

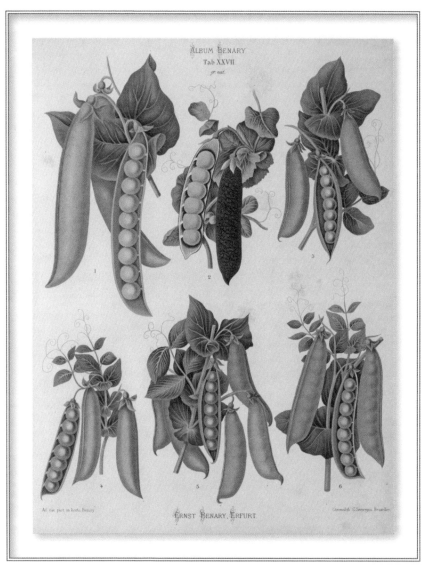

'Peas', by Ernst Benary (1819–1893)

from Album Benary

COLOUR LITHOGRAPH BY G. SEVEREYNS

TRADITIONAL STORAGE METHODS

Prior to modern conveniences, such as canning, freezing and shipping foodstuffs by air freight around the world, fruit and vegetables had to be stored to provide food throughout the winter months. Root vegetables, such as carrots, potatoes and parsnips, were completely buried in sand and kept in dry, cool storerooms (a method still employed by gardeners today). Leafy types, such as cabbages and lettuce, were harvested with their roots still intact, then partially buried in indoor trays of soil to preserve freshness.

More elaborate and time-consuming techniques now make reaching for the can opener seem something of a luxury! One such method was the traditional earth-covered 'clamp', used for storing large quantities of potatoes. Sited in an out-of-the-way corner of the kitchen garden, a thick bed of clean straw was laid on the ground and the vegetables piled on top, forming a long ridge. Straw was placed on top, then another layer of potatoes added, then more straw. Spaces were left for ventilation to ensure the tubers did not begin to 'sweat'. Finally, all was covered with a thick layer of earth. This was partially removed each time a new supply of produce was needed, then carefully resealed. A channel was dug around the clamp to catch the rain and keep the mound dry. The same method was used for storing carrots. However, the circular carrot clamp had a characteristic igloo shape, formed by placing the carrots with their leaf ends pointing outwards.

SOME PARSNIP FACTS

The sweet parsnip (*Pastinaca sativa*) was prized in medieval times as a food and medicine. It was thought to alleviate toothache, stomach pains and impotence. In culinary terms it has been somewhat superseded by the carrot. In Italy pigs are fed parsnips in the belief that it enhances the flavour of *prosciutto* (dry cured ham). Rich in carotenoids, vitamin C, calcium, potassium, and fibre, its flavour is improved considerably if lifted only after the first frosts (when the starch is turned to sugar, making them sweeter). Old cultivars include 'Hollow Crown' (also known as 'Long Jersey') and the 1897 'Tender and True', said to be named after a popular song of the time!

FORGOTTEN VEGETABLES

Given its famed reputation for all things gastronomic, it is perhaps not too surprising to find that France is home to some of the most passionate and innovative vegetable growers in the world. While in Britain there are many fans of heritage seed, and Americans favour heirloom varieties, the French enthusiastically revive *les legumes oubliés*, the forgotten vegetables.

HOW TO SOW HEIRLOOM SEEDS

Method for sowing seeds in containers for indoor cultivation.

1. Fill a seed tray or pot with fresh seed compost and level to a half an inch below the rim. Never use old compost saved from the previous year because this can adversely affect seed germination as well as subsequent plant development.
2. If using fine seeds carefully sprinkle them thinly over the whole surface as evenly as possible.
3. Gently sift the compost over the whole surface, aiming to cover the seeds with no more than their own thickness in compost.
4. If sowing larger seeds make an indent in the seed compost at the required depth with the end of a pencil and drop in a single seed. Cover the seed holes with sifted compost.
5. Water sparingly, using a watering can fitted with a fine 'rose' attachment. Overwatering can cause seeds to rot. Cover with a plastic or glass sheeting and keep out of direct sunlight.
6. Label with the plant name, variety and date.
7. Check daily and, as soon as seedlings begin to appear, remove the covering and keep lightly watered, but be careful never to drench with water.

Method for sowing seeds in furrows outdoors.

1. Ensure the area is free from weeds and has been raked to produce a fine soil texture and has a flat and even surface.
2. Use a piece of string pulled taut between two stakes or lay a stake across the soil as a sowing guide. Make a shallow furrow (about 2cm/1in deep) along the length of the marker with a hoe.
3. Scatter the seeds as evenly as possible along the length of the furrow. Larger seeds can be individually dropped into the furrow at the desired distance.
4. Gently rake the soil over the furrow.
5. Water sparingly, using a watering can fitted with a fine 'rose' attachment.
6. Place a label at the end of the row with the plant name, variety and date.

As seedlings develop, keep the area watered and free from weeds.

THE CAULIFLOWER

Brassica oleracea var. botrytis

BRASSICACEAE

IT IS BELIEVED the cauliflower originated in the Middle East and it was well known in Europe by the Middle Ages. The English herbalist John Gerard (1542–1611), in his *Herball* of 1597, rightly advised that the 'colieflore' should be sown on a hot dung pile during early spring. However, like the Brussels sprout, the cauliflower was not consumed in great quantities in the United States until the 1920s.

Cauliflowers are available with either white, pale green or purple heads (also known as the curd). 'Romanesco' is a highly decorative cultivar with pointed florets of a striking yellowy-green colour. Occasionally this is sold as calabrese, but is really a cauliflower. There are many heirloom varieties for the home grower to choose from, including 'Early Snowball' and 'Perkin's Leamington'. The splendidly named 'Veitch's Self Protecting' (from the 19th century) has outer leaves that curl tightly around the head, protecting it until it reaches full maturity. Cauliflowers are not a good vegetable for the novice gardener, although they are easier to grow successfully throughout the cooler winter months. They can be eaten raw or lightly cooked. Alkaline water is said to turn cauliflowers yellow; to prevent this from happening add a little milk or lemon juice to the water. Aluminium cooking vessels should be avoided for the same reason, but iron pots result in unappetizing brownish-blue cauliflowers!

Brassica oleracea var. botrytis

'Parisian cauliflower', colour lithograph by Elisa Champin (n.d.)
from Album Vilmorin
VILMORIN-ANDRIEUX & CIE (1850–1895)

NAME THAT VEGETABLE QUIZ

Heirloom vegetables often have very descriptive names, such as the cabbage 'Large Drumhead Savoy', but others give no clue at all as to what type of vegetable they identify! Here are a few to test your vegetable knowledge – see if you can identify what they are. (Answers at bottom of the page.)

1. 'Walla Walla'. . . . .
2. 'Kentucky Field Cheese'. . . . .
3. 'Miniature Chocolate Bell' . . . .
4. 'Black Gnome'. . . . .
5. 'Grandpa Admire'. . . . .

DIG FOR VICTORY

The 'Dig For Victory' slogan was introduced in Britain by the Ministry of Defence barely a month after the outbreak of war in 1939. A revival of the First World War campaign to encourage civilians to grow their own food, its results exceeded all expectations: the response was so enthusiastic that food imports into Britain were effectively halved. Not only did back gardens and allotments begin to resemble smallholdings but public spaces were also turned over to the cause. London's Hyde Park boasted its own piggery while the herbaceous borders of Kensington Gardens sprouted cabbages and potatoes where roses and carnations had once grown.

The Ministry of Food sought to enliven what must at times have been a somewhat relentless diet of greens and roots, with such graphic promotional characters as Doctor Carrot (with his interesting sounding carrot jam) and Potato Pete who advised that potatoes should be scrubbed, not peeled, thus preserving nutrients and avoiding waste. As she scrubbed away the patriotic housewife would recite:

Those who have the will to win,
Cook potatoes in their skin,
Knowing that the sight of peelings,
Deeply hurts Lord Woolton's feelings.

Lord Woolton was the Minister of Food and the inspiration behind the eponymous Woolton Pie. If a little lacking in culinary treats, the wartime diet – rich in home-reared meat, eggs and vegetables – is now recognized as being the healthiest ever enjoyed by the British public.

ANSWERS

1. An onion from the French island of Corsica. 2. A squash. 3. A small, green, sweet pepper that turns brown as it ripens. 4. A small, black aubergine. 5. A very old, looseleaf lettuce.

HOW TO PROTECT HEIRLOOM VEGETABLES
FROM SLUGS & SNAILS

There seems no limit to the inventiveness (or is it just desperation?) of gardeners when adding to their arsenal of weapons deployed against the humble slug and snail. The following are just some of the techniques employed to date that may be worth trying.

1. Paths drenched with an infusion of walnut leaves, salt and crushed eggshell were once thought to deter the nighttime wanderings of slugs and snails.

2. Crushed eggshells are often used in an attempt to create a defence around vulnerable plants. A more sophisticated version is the zinc collar. Shaker gardeners would sprinkle sand or grit around their plants to act as what they termed a 'modesty border', because slugs and snails find both substances difficult to traverse.

3. Allowing domestic fowl, such as ducks and chickens, to pick over the kitchen garden can be an efficient way to clear an area of these pests. However, experience shows they are not always 100 per cent reliable at discriminating between precious seedlings and slimy slugs.

4. Compassionate gardeners thoughtfully gather up the offenders, place them in a bucket, then transport them well away from their vegetable patch to areas of rough grass. However, experiments in which the shells of snails have been marked show that they often return to their home patch, sometimes travelling surprisingly long distances to do so.

5. Slugs and snails are enthusiastic beer drinkers, so some gardeners sink shallow containers of cheap beer among their leafy vegetables, in the hope that the greedy culprits will fall in and drink themselves to death.

6. If all else fails, the desperate gardener can always choose to embrace their unwelcome harvest and eat the meat of the snail, as the French do. Advocates of certain types of traditional African medicine make claims for the benefits of consuming snail meat. Singers say it keeps them in good voice and, when used as an ingredient in 'love medicine', it is said to restore marital bliss!

THE BEAN

Phaseolus vulgaris & *Vicia faba*

F A B A C E A E

REMAINS OF THE BEAN *Phaseolus vulgaris* have been found in Peruvian caves dating back to 6000 BCE. Originating from Mexico and Central America, Native Americans have eaten them for thousands of years and the climbing bean is a constituent part of their culture's totemic 'Three Sisters', the system of growing corn, beans and squash plants together (the corn acts as a support for the beans). Sometimes referred to as the common bean, the *Phaseolus* group of beans contains the French and haricot types. Climbing and bush types belong in this category and both the bean and pod are eaten. This group of beans was introduced from North America to the rest of the world in the 16th century. It quickly became popular in Europe, but initially only the climbers were eaten. It was not until the 18th century that the bush varieties were in wide cultivation. Along with the ubiquitous green bean, other colours available to the home grower include purple, pink, red, white, yellow and even variegated cultivars.

When preparing the beans of the broad bean type, *Vicia faba*, only the shelled bean is eaten – the pod is discarded. Originating in the Mediterranean region, these are some of the earliest plants cultivated by humankind and they were taken to the New World in the 16th century by the Spanish colonists. Both climbing and bush varieties are available.

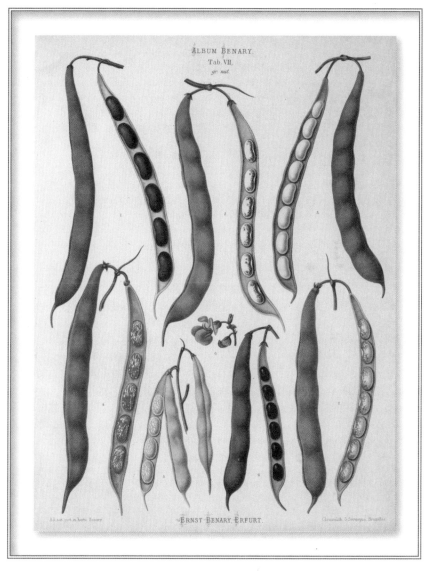

'Bush beans', by Ernst Benary (1819–1893)

from Album Benary

COLOUR LITHOGRAPH BY G. SEVEREYNS

SOME ASPARAGUS FACTS

That most delicious of vegetables, asparagus (*Asparagus officinalis*) is also known as coralwort due to its red berries. Originally, it was the thin green type that was introduced into England by the Romans and it was not until the 17th century that the much larger, fatter, white Dutch asparagus reached the plates of wealthy aristocrats. Asparagus has traditionally been a delicacy only enjoyed by the rich. It needs a great deal of space to grow successfully and produces spears for only a very short season. To the poor, the yield would seem disproportionately small to the space allotted to the plant.

Other, unrelated, edible plants also bear the name – for example asparagus chicory. Popular in Italy, where it is affectionately known as *puntarella*, the tips and leaves are used in salads or lightly cooked and dressed with lemon juice and olive oil. Many Italian-Americans refer to it as pine cone chicory. Celtuce (*Lactuca sativa* var. *asparagina*), sometimes called asparagus lettuce, has been grown in China for centuries, and came to North America in the 1890s. It is cultivated for its thick, round, central stem, instead of its leaves. The stem can be peeled and boiled briefly, or eaten raw. By contrast, it is the young, whole seedpods of the winged pea (*Lotus tetragonolobus*) that are eaten. It goes by the alternative names of asparagus pea or Winged Bird's-foot Trefoil.

A BRIEF HISTORY OF LETTUCE

Representations of lettuces appear in Egyptian tomb paintings dating back to 2700 BCE. The Romans cultivated several varieties, including the types we know as romaines, cos and butterheads. They ate raw young lettuce and cooked or wilted older ones in hot olive oil and vinegar. Lettuce is mentioned in various ancient texts.

HALLOWEEN LANTERNS

Each year on the eve of All Hallows, 31st October, thousands of orange pumpkins are sacrificed in a frenzy of lantern making, as the tough vegetables are hollowed out and carved before a flickering candle is placed inside. The origins of this tradition are somewhat confusing, with myriad variations of the story told. The general thrust of the tale begins with an Irish farmer, known as Stingy Jack, who upset both God and the devil. Barred from entering heaven, Jack bargained with the devil not to take his soul. Destined to wander the earth in darkness for eternity, Jack was tossed one of hell's everlasting glowing embers by the devil. He carved out a turnip and put the ember inside to act as a lantern to light his lonely way. Once the tradition was transported to North America, the humble turnip was superseded by the more glamorous pumpkin, now known as jack- o'-lantern. In fact 'jack-o-lantern' originally referred to a man with a lantern or a night watchman.

SOME ONION FACTS

ༀ

The ancient Egyptians revered the
onion, regarding its spherical shape and concentric
rings as symbols of eternity. It was also regarded as an
aphrodisiac, making it a forbidden fruit for Egyptian priests, for fear
that it would inflame their desires. Many classic Hindu texts on
lovemaking also cite its efficacy in this area. In traditional French households
onion soup was served to newly-weds for breakfast following their wedding night.
Christopher Columbus included the onion among his supplies on his
first voyage to North America in 1492. The benefits of this enduring vegetable
would not only have been appreciated in the galley but also in the sick room
(had there been such a thing aboard the cramped ship) as its medicinal properties
are legendary. These include alleviating the symptoms of sore throats, colds and
flu; calming skin inflammations such as boils, blisters, cuts and insect
stings; restoring circulation to frozen feet by rubbing them with raw
onion; and slowing bone deficiency disorders. Libyans
consume more onions than any other nation, eating
well over 30kg/60lb per head per year!

ༀ

A BRIEF HISTORY OF THE AUBERGINE

Solanum melongena, the aubergine, known as the eggplant in the United States, was
reported from Aleppo, Syria, in 1575 by Leonhard Rauwolf. It is an ancient vegetable first
recorded in India (where it is called *brinjal*) and it has several Sanskrit names. The first type
to arrive in 16th-century England had white fruits the size of a hen's egg, hence the
American name. Rumour has it that Arab brides were once required to know at least a
hundred different ways to prepare an aubergine!

There are many varieties that differ considerably in shape and colour from the more
common purple and white kinds, including green, pink, yellow and variegated ones. The
home grower is somewhat spoiled for choice when it comes to selecting which aubergine
to cultivate. Here are a few: 'Apple Green' dates back only to the 1960s and has pale green
round fruits. 'Imperial Black Beauty' has oval fruits and is the result of crossing 'Large Early
Purple' and 'Black Peking' (a cultivar that arrived in the United States from China in 1866).
'Indonesian Pink Blush' is tinged pink and violet. With its oval fruits and white skin
streaked with bright pink, 'Listada da Gandia' originated in Italy, then was introduced into
France in the mid-19th century, where it was known as 'Striped Guadaloupe'.

THE RADISH

Raphanus sativus

BRASSICACEAE

RADISHES HAVE A LONG HISTORY, appearing in Egyptian hieroglyphics and drawings at the temple of Karnak – indeed they helped fuel the diet of the thousands of workers who built the pyramids! The Chinese greatly prize the radish for its medicinal qualities and ferment them with other plants in hermetically sealed ceramic pots. This process can take up to forty years and the resulting preparation is used to treat a range of problems, including fevers, stomach disorders, intestinal infections, ulcers and flatulence.

The range of cultivars available to home growers is considerable, with forms that include round and turnip-shaped ones, as well as those with long, tapering roots. As the names 'Long White Icicle' and 'Purple Plum' suggest, radishes come in an array of different colours apart from the red ones so commonly available. Look out for seeds of black, pink, purple, yellow and white varieties to grow. Some, such as 'French Breakfast', are bicoloured, with this type being scarlet at the top fading down to white.

The best crops of radishes are achieved by sowing in well-prepared soil, with generous and frequent watering and reasonably consistent temperatures. They should be harvested as soon as they are ready, and not left in the ground too long or they will become tough and woody. Sow successionally for a continuous supply of fresh, tasty and crunchy jewel-like radishes.

'Radishes', by Ernst Benary (1819–1893)

from Album Benary

COLOUR LITHOGRAPH BY G. SEVEREYNS

THE CARROT
Daucus carota

APIACEACE

THE EARLIEST WILD CARROTS probably originated in Afghanistan and would have been deep red or purple in colour. Certainly the ancient Greeks and Romans cultivated violet Syrian carrots for their medicinal, as well as culinary, properties. Alexander the Great brought little, round, green carrots back from India, and purple and yellow cultivars were recorded in 13th-century China and later in 18th-century Japan. However, they only reached North America when the English colonists planted them in their Virginian kitchen gardens in 1609. Carrots have not always been thought of as simply a vegetable – in about 1300 Italians were eating violet cultivars sweetened with honey as a dessert, and in the 15th century aristocratic English ladies adorned their hats with the feathery foliage of carrot plants.

Heirloom growers can choose from an array of variously shaped carrots in colours that range from deep red and purple to yellow, white and orange. The violet types can have a yellow core and are sometimes referred to as the black carrot. It is said that the now familiar orange carrot was originally bred by 17th-century Dutch nurserymen as a patriotic act, because orange was the national colour of their independent state! Certainly the progenitors of many of the carrots we eat today were bred in the Dutch town of Hoorn and have names like 'Early Scarlet Horn', 'Early Half Long', and 'Late Half Long.'

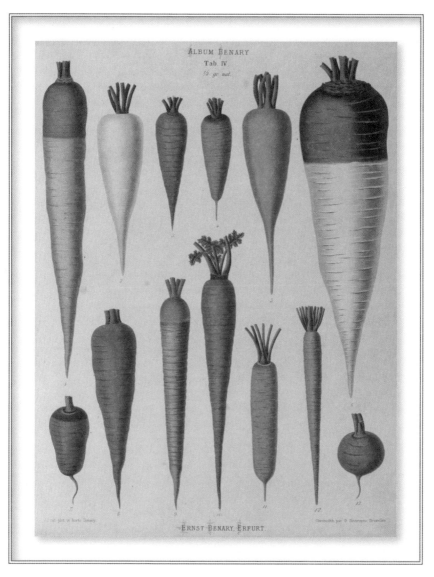

'Carrots', by Ernst Benary (1819–1893)
from Album Benary
COLOUR LITHOGRAPH BY G. SEVEREYNS

RADICCHIO

Radicchio, also referred to as red chicory, is a highly decorative leaf vegetable with a tight heart at its centre. Either the leaves are used raw in salads or the heart is cooked. It is beautiful enough to steal the show growing in any *potager* and adds glamour and colour to the salad bowl. The 18th-century Italian cultivar 'Castelfranco' has lovely mottled green, red and creamy-white leaves. 'Red Treviso' is even older, dating from the 16th century. In summer the leaves are green but, as temperatures drop and winter approaches, it turns bright red with white veining. Some people are put off by its bitterness, but it makes a great salad when teamed with oranges and red onions to add sweetness.

PARTICULAR PESTS

Pests are an ever-present responsibility for all home growers. It appears some small enemies are none too discriminating and seem happy enough to devour most things in the vegetable garden. However, a brief look at the following list shows that many pests actually have very specific tastes!

Asparagus beetle	*Crioceris asparagi* – enjoys asparagus.
Black bean aphid	*Aphis fabae* – loves beans and peas.
Cabbage looper	*Trichoplusia ni* – lettuce and cabbage are their preferred diet, but they will eat most things!
Cabbage root fly	*Delia radicum* – munches on all brassicas.
Carrot root fly	*Psila rosae* – eats celery, celeriac and parsnips.
Celery worm	a.k.a. parsley worm or carrot worm *Papilio polyxenes* – fond of dill and parsnips.
Colorado potato beetle	*Leptinotarsa decemlineata* – loves potatoes.
Pea moth	*Cydia nigricana* – peas, and moves on to beans once the peas are gone.
Spotted cucumber beetle	*Diabrotica undecimpunctata* – partial to beans, asparagus, tomatoes and cabbages.
Turnip sawfly	*Athalia rosae* – loves turnips.

TOMATO TASTE

In tomato taste tests conducted in Britain, red varieties scored the highest points over other colours. However, when the testers were blindfolded, orange varieties triumphed! Among some of the best orange cultivars are the globe-shape 'Gold Dust', the golfball-size 'Livingston's Golden Ball', and the unusual, fuzzy skinned 'Garden Peach'.

CABBAGE SUPERSTITIONS

Several superstitions are attached to the humble cabbage. If it is eaten on New Year's Day, good luck should be guaranteed for the coming year, especially if black-eyed peas feature in the meal too. The leaves of the plant hold the promise of prosperity, as they are symbolic of bank notes. Most bizarre of all is the old Halloween practice of sending out into a cabbage field a pair of blindfolded girls. It is said that the condition of the first cabbage they find and pull will predict much about their future husbands. If plenty of earth is attached to the roots, their future suitors will be rich; if bare-rooted, then poverty awaits them; then, once cooked, whether or not the cabbage tastes sweet or sour is indicative of their husbands' dispositions.

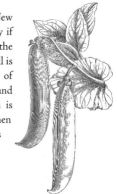

PEA & BEAN KNOW-HOW

As a novice vegetable grower, it is all too easy to become confused by the various types and names for beans and peas, especially when choosing seeds. Here is a brief explanation.

Peas	*Pisum sativum* – including garden peas, snow peas, and snap peas. Both tall and bush cultivars are available. Peas and garden peas are shelled while snow peas and snap peas are usually eaten whole (either raw or lightly cooked), although some varieties can be left to develop proper peas and then shelled.
Broad beans	*Vicia faba* – also known as field, fava, and bell beans. Tall or short cultivars are available and only the shelled beans are eaten.
Butter beans	*Phaseolus lunatus* – also known as lima beans, either grown as a climber or bush and needs the heat of its native South America to grow well. Only the beans are eaten.
Black-eyed peas	*Vigna unguiculata* – Several species belong to the genus *Vigna*, known by their common names as crowder, cowpea, southern and yard-long beans. The pods can be cream, green, pink or purple with the beans varying in colour, too. They need a hot climate to thrive and only the beans are eaten.
Runner beans	*Phaseolus coccineus* – these are tall plants that need support, and the whole pod is sliced and eaten.
French & haricot beans	*Phaseolus vulgaris* – both the climbing and bush types are smaller, more delicate than runners, and can be green, purple or yellow. Old types include the Dutch cultivar 'Dragon Tongue', which is speckled cream and purple. French are eaten as green beans; haricots can be eaten fresh (flageolets) or dried (haricots).

THE CELERIAC

Apium graveolens var. *rapaceum*

A P I A C E A E

LIKE CELERY, celeriac is a descendant of the coastal-loving wild celery plant. This plant was used as a herb, much like parsley, and its seeds as condiment. Indeed celery salt (salt mixed with celery seeds) is much valued by cooks today. The bulbous-rooted celeriac is thought to have developed as a variant of stalked celery in the eastern Mediterranean region. It first appears in seed catalogues in the early 18th century and it is widely used throughout Europe. The edible part of the plant is not, in fact, the root itself but the swollen lower stems. Celeriac foliage is crisp and tasty – an added bonus for those who grow their own.

It has to be admitted that celeriac would win no beauty contests, and this may account for its lack of popularity in some countries, especially North America (where it is sometimes called celery root). With its knobbly, brown-skinned, round, swollen stem it actually resembles a turnip rather than its slender, long-stalked cousin. Not surprisingly, one of its common names is turnip-rooted celery, but it is also known as German celery, an acknowledgment that the country is rightly enthusiastic about this vegetable and fully appreciates its hidden qualities! Celeriac is far sweeter than celery with a delicate and nutty flavour. It can be grated finely and used in salads and makes wonderful soups and stews and is delicious mashed. A 19th-century cultivar, 'Early Purple Vienna', has small bulbs with purple stems.

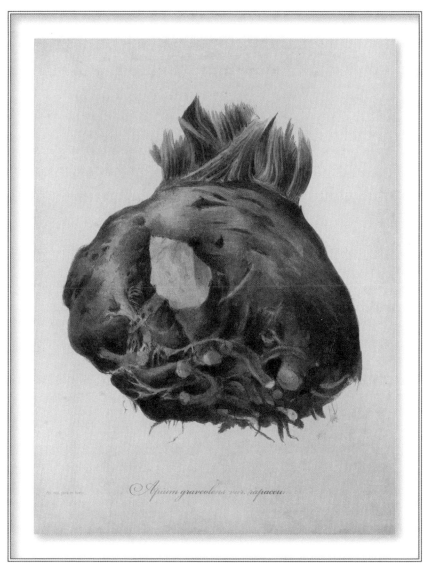

'Celeriac', colour lithograph by Elisa Champin (n.d.)
from Album Vilmorin
VILMORIN-ANDRIEUX & CIE (1850–1895)

THE LETTUCE
Lactuca sativa

ASTERACEAE

ALTHOUGH NATIVE TO EUROPE and the Mediterranean regions, the first depiction of the lettuce is believed to occur on an Egyptian tomb relief. Dating back to the third century BCE, it shows the cultivation of a tall romaine-like plant. Lettuce is known to have been eaten by the ancient Greeks and widely cultivated by the Romans, who ate it after their main course (a tradition that remains popular in Italy today).

Although it is the ubiquitous basis for countless salads, inventive cooks and adventurous gardeners need never feel the need to resort to the characterless iceberg type because the forms, colours, and textures of the lettuce are numerous. Those known as the heading or crisping lettuce include the batavia and butterhead types. The former (of which the iceberg is one) has crisp leaves and ranges in colour from yellow to dark green. The latter type has smooth leaves with a slightly jagged edge and is much prized for its soft texture and buttery flavour. The long-leaved romaine type originates from the Greek island of Cos (which is why it is also called the cos lettuce) and is loose headed, with generous, upright, crunchy leaves. The cutting group of lettuce, also referred to as 'cut-and-come-again', form no heart and are very popular with gardeners because they can be harvested by cutting to just above ground level, and then they will sprout fresh and plentiful new growth.

'Round lettuce', by Ernst Benary (1819–1893)

from Album Benary

COLOUR LITHOGRAPH BY G. SEVEREYNS

RADISHES

Traditional herbalists used the radish (*Raphanus sativus*) in a variety of preparations. Kidney stones would dissolve, as if by magic, when the fresh juice of the radish root was taken with a little white wine. The English botanist Nicholas Culpeper (1616–54) advised: 'The roots eaten plentifully sweeten the blood and juices, and are good against the scurvy.' They were also thought to treat malfunctions of the bladder and, if harvested as the moon waned, would cure corns and warts. More recently, Professors Esch and Gurrusiddiah, from the State University of Washington, extolled the antibiotic properties of the radish.

It is now widely recognized that eating generous portions of brassica vegetables (to which family the radish belongs) helps protect against many cancers. Sadly the radish is one of those vegetables that is all too easily overlooked when shopping. After all, they can just seem like so many small, round, red crispy balls, but the reality is very different. Many heirloom cultivars are still available and several certainly appear to be worth growing. 'Black Spanish Round' dates back to the 16th century and has black skin and pure white flesh. The delightful-sounding 'China Rose' has a long form with rose-coloured skin, and was introduced into Europe from China by Jesuit missionaries in the mid-19th century. 'French Golden' is, as its name suggests, a lovely light golden colour and 'German Beer' is white-skinned, while 'Purple Plum' is bright mauve.

A BRIEF HISTORY OF PUMPKINS & SQUASHES

Pumpkin seeds have been found in Mexican caves believed to date back to as long ago as 7000 BCE. Certainly pumpkins and squashes were grown and consumed in sixth-century Africa, China and India, but not introduced into Europe until the 16th century, when the vegetables were more highly prized for their use in sweet, not savory, dishes. The term 'pumpkin' dates from the 17th century, and comes from the Greek for melon, *pepon*, meaning 'cooked by the sun'.

Over time the medicinal uses for pumpkin and squash have included the treatment of such unmentionable conditions as tapeworms, boils and warts.

TOMATOES & ARTHRITIS

People who have arthritis are often advised to avoid tomatoes because of their high levels of acidity. However, all is not lost as the white, yellow, orange and pink cultivars are thought to be far kinder to those with the condition.

HOW TO GROW HEIRLOOM LETTUCE

Many heirloom cultivars of lettuce fall into the category that used to be known as cutting lettuce, which are now more commonly referred to as the looseleaf or 'cut-and-come-again' types. 'Bronze Arrow', 'Deer Tongue' and 'Red Oak Leaf' are all examples of cutting lettuce. These do not store or travel well, so are often best grown at home. Apart from the obvious benefits of freshness and taste, the sheer visual pleasures provided by the variety of colour and leaf shape of these heirlooms is reason enough to grow them.

1. To get an early start, sow lettuce seeds indoors, then transplant the young plants into their final growing position as conditions warm up. However, they dislike disturbance so only plant them when a strong root system has developed and there is plenty of soil attached to the roots.

2. If planting directly into the ground, ensure the patch is well weeded and fertilized and that the soil is free from clumps and stones.

3. Lettuce seeds are very fine so plant them shallowly and as evenly spaced as possible. Firm the soil, then water using a watering can fitted with a very fine 'rose' attachment to avoid washing the seeds away.

4. Lettuces do not like too much strong sun. Provide shade by interplanting them at the base of tall-growing plants, such as sweetcorn, or between rows of climbing beans. Alternatively, plant closely in blocks rather than rows, because this helps conserve moisture in the soil.

5. Young leaves are sweet and tender, while lettuce that have been in the ground too long or have gone to seed (flowered) will be tough and bitter-tasting.

6. If possible, harvest the leaves early in the morning before the heat of the day has had a chance to wilt them.

7. To avoid feasts and famines, organized gardeners make regular sowings of lettuce throughout the growing season. Successive sowings every three weeks or so should provide a continuous supply of fresh and tender leaves.

THE ONION

Allium cepa

ALLIACEAE

ONION REMAINS have been unearthed in Neolithic Age settlements in Jericho, Palestine, dating back to 5000 BCE. Thought to have been native to Afghanistan, Pakistan and Iran, the onion was a staple of the Egyptian diet around 3000 BCE. Seeds were found in Egyptian tombs and the plant features in relief carvings. The onion was introduced into India and many Mediterranean regions via early trade routes, and certainly by Roman times its use was widespread.

The onion is now an indispensable ingredient in everyday cooking. Just consider how many recipes begin with the phrase 'peel, slice and sauté an onion until soft.' Ranging in colour from yellow and white to red, they may be round, slightly flattened, or teardrop in shape. Spring and early summer cultivars are used fresh while the late summer and fall cultivars store well and are used throughout the winter months.

Although onions are commonly available in grocers and supermarkets, many people prefer to grow heirloom cultivars as they offer a greater range of uses, with flavours ranging from types that are very sweet and suited to eating raw, to strong and sharp onions only suited to cooking. The intriguingly named 'Walla Walla Sweet' onion originated on the French island of Corsica and was brought to 'Walla Walla' in Washington State in the early 19th century by Peter Pieri, a French soldier. It is a large, globe-shaped onion favored for its sweet and mild flavour.

'Onions,' by Ernst Benary (1819–1893)

from Album Benary

COLOUR LITHOGRAPH BY G. SEVEREYNS

SALSIFY & SCORZONERA

Bored with the humble carrot? Then why not try these unusual roots as an alternative? Salsify (*Tragopogon porrifolius*) also goes by the names vegetable oyster and oyster plant. Its long white roots resemble a thin parsnip and have a very delicate flavour. An 1897 American seed catalogue gives the following description: 'The Salsify is a hardy biennial plant, and is principally grown for its roots, which are long and tapering, and, when grown in good soil, measure twelve or fourteen inches in length. It is considered wholesome and nutritious. When cooked, the flavour resembles that of the oyster, and is a good substitute for it: hence the name.' Scorzonera (*Scorzonera hispanica*) is similar to salsify but has black roots, and some claim that of the two it has the superior taste. Also known as the black salsify.

VEGETABLE PHOBIAS

If any particularly precocious vegetable-hating youngsters read this, they may decide to cite medical opinion to justify all future refusals of broccoli at the dinner table. Lachanophobia is the fear of vegetables. Close encounters of the vegetable kind can induce in people with this phobia rapid breathing, an irregular heartbeat, nausea and sweating. More rarefied versions include lachanophobia mycosis (fear of mushrooms) and lachanophobia lycopersicum (fear of tomatoes).

A BRIEF HISTORY OF CAULIFLOWERS

The humble cauliflower (*Brassica oleracea* var. *botrytis*) is both beautiful to look at and wonderful to eat. It hails from the Middle East and has been cultivated in Europe from the 13th century. Its old English name is 'coleflower' or 'cabbage flower'. Americans did not begin to eat the vegetable in any great quantity until the 1920s. The author Mark Twain (1835–1910) once famously referred to the cauliflower as a 'cabbage with a college education'! The firm white flower head is composed of smaller bunches of flowers known as florets or curds.

Some cooks think the whiter the flesh of the head, the better the taste, while others prefer those whose florets are edged with green. As its name suggests, 'Purple Cape' is a colourful cultivar that is very sweet tasting. It was known in England as early as 1808 and may originate from South Africa. Whatever the colour of a cauliflower, it should never be overcooked.

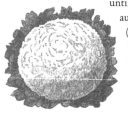

HOW TO GROW & HARVEST HEIRLOOM ONIONS

Onions have long been one of the main staple crops grown by commercial and amateur gardeners alike. Here are a few traditional tips for cultivation, harvest and storage.

1. Today, onions are grown either from seeds or from sets (the immature small bulb). The home grower will find a far wider range of cultivars available as seeds.

2. For large bulbs a rich loamy soil is preferable. Those grown in light and sandy soil are more liable to attack by pests, such as the dreaded onion fly, which lays its eggs in the bulb. The hatched maggots then munch away on the flesh.

3. A nearby planting of parsley is thought to provide a helpful deterrent against pests.

4. Traditionally, main crops were sown as early in the year as possible, as soon as the ground was workable. A later sowing in midsummer then ensured a fresh supply of onions during winter.

5. Once the leaves of onions have flopped over, leave the bulbs undisturbed in the ground to mature for around two weeks before harvesting. Gently lift the bulbs from the soil, preferably on a dry, sunny day, then leave outdoors for a few days to cure. Cut away the leaves to a couple of inches from the bulb. Leave in a warm but not humid place indoors for two or three weeks, turning regularly to ensure even drying. Allow plenty of space between the bulbs so air can circulate freely. Store in hanging nets in a cool, dry place.

6. In the 19th century these were considered some of the best onion cultivars: 'Brown Globe', a hardy onion and useful for an early crop; 'Giant Madeira', which, as its name suggests, grew to a great size and had a mild flavour; the small 'Silver-Skinned' was favoured for pickling, while 'White Lisbon' was sown in fall and picked as a scallion, and is still a popular cultivar today.

THE SWEETCORN
Zea mays

SWEETCORN IS A NATIVE of the Americas and is a plant of great antiquity. No record of it survives elsewhere prior to its introduction into Europe by Christopher Columbus. The yellow cultivars of sweetcorn are the most readily available today, although the home grower will easily be able to procure the seeds of old cultivars that will produce beautiful rainbow-coloured cobs! Blue, orange, pink or white kernels all exist as well as wonderful multi-coloured cultivars such as 'Cocopah', which was collected by gold prospectors from Colorado River Native Americans in 1868.

Despite this array of colours, in the past many Americans only favoured the white cultivars of corn, while the yellow types were thought more suited for horse fodder than the dinner table. This all changed at the turn of the 20th century when seeds of the heirloom variety 'Golden Bantam' became commercially available. Indeed, in 1926 the Burpee Seed Company proclaimed 'Golden Bantam' as 'America's Favorite Sweet Corn', and this variety is still widely grown today.

Sweetcorn should be cooked as quickly after harvesting as possible. Mostly it is plunged briefly into a large pan of boiling water and, until the 20th century, it was usually cooked in the husk to better preserve the flavour. The cooking water should not be salted because this hardens the kernels. Sweetcorn cobs are also excellent grilled or barbecued.

'Sweetcorn', colour copper engraving by Joseph Jakob von Plenck (1738–1807)
from Icones Plantarum Medicinalium
R. GRAEFFER, VIENNA (1788–1803)

MUSHROOM SUPERSTITIONS

It is not that uncommon to find mushrooms growing in a ring formation in fields and woodlands. These are known as fairy rings. Although these rings occur naturally (formed by the fruit bodies around the edge of a circular mycelium) and increase in size each year (as the fungus spreads), numerous superstitions and myths have become associated with them over time. Their existence has variously been attributed to the 'little people', lightning strikes, shooting stars, even meteorite showers. Although it is often thought that treasure is buried within the ring, those who enter risk becoming either blind or lame or disappearing altogether underground. In Scottish mythology the large puffball is known as 'the devil's snuffbox' because its spores were believed to cause blindness. More positively, medicinal concoctions made from mushrooms are believed to ripen boils and abscesses!

TUTANKHAMEN'S PEAS

Pea seeds were among the many treasures found by archaeologist Howard Carter and his patron Lord Carnarvon in 1922 when they excavated the tomb of the famous Egyptian boy king Tutankhamen. (Garlic bulbs were also discovered in the tomb.) Today, growers of heritage cultivars can grow seeds of the eponymous Tutankhamen pea, which is thought to originate from Carnarvon's estate at Highclere Castle, Berkshire, England.

RHUBARB

Technically speaking, rhubarb (*Rheum rhaponticum*) is actually a vegetable, although most people consider it to be a fruit and consequently use it in sweet dishes, such as pies, or in preserves. It is an ancient vegetable and probably originates from Tibet. Today, it is grown as a perennial plant in many gardens. It is thought to have been introduced into North America around 1800 by a farmer in Maine. The long pink stalks are the edible part of the plant while the leaves should never be eaten as they contain enough oxalic acid to cause death. Rhubarb is served as a tart accompaniment to potatoes in some Polish recipes.

COWS & CARROTS

In the 17th century, Dutch cows were fed on carrots, because this diet produced the very richest milk and the best colour in butter.

Elsewhere, the milk of less well-nourished bovines was mixed with carrot juice, thus improving its colour.

HOW TO GROW HEIRLOOM SWEETCORN

Sweetcorn is pollinated by the wind instead of by visiting insects. If you want to save your own seeds, it is advisable to grow just one cultivar. Alternatively, if you have the room, you can space the various cultivars well apart (a minimum of 8m/26ft is advised), or even place paper bags over the corn ears to stop cross-fertilization.

1. Either start the sweetcorn indoors, sowing the seeds 4cm/1.5in deep into individual pots or, after all danger of frost has passed and once the soil has warmed a little, sow seeds directly into the prepared ground.
2. When plants are 15cm/6in high, space them 30cm/12in apart and plant in blocks measuring 2m x 2m/6ft x 6ft. Directly sown plants should be thinned to this distance. Do not plant in rows. Planting in blocks helps with pollination.
3. Water well and continue to do so in dry periods. As the sweetcorn tassels begin to emerge from the stalk, they need plenty of water for the kernels to develop, so ensure the ground does not dry out, especially at this stage.
4. The kernels should be ready to harvest about three weeks after the first silky strands appear. As the tassels turn brown, peel a little of the leaf away and carefully pierce the corn with a fingernail; if the released liquid is milky, it is ripe. If not, leave on the plant a little longer.
5. Like asparagus, sweetcorn should be cooked and eaten as soon after harvesting as possible because the sugars quickly begin to turn to starch.

Many heirloom cultivars are available, a number of them from America. Among some of the more unusual are 'McCormack's Blue Giant', which produces smoky blue kernels, and 'Black Mexican', the kernels of which are white to begin with but dry to black. The delightfully named 'Texas Honey June' is so sweet-tasting that it has been likened to honey.

THE GARLIC

Allium sativum

ALLIACEAE

GARLIC IS THOUGHT to be a native plant of Central Asia and is known to have been used by the Chinese and Egyptians thousands of years ago. Its medicinal and mythological qualities are legendary. The ancient Egyptians, Greeks and Romans all associated the consumption of garlic with greatly increased strength and courage, feeding it to their slaves and gladiators. It is now a vital ingredient in many cuisines around the world.

It is the swollen bulb of the plant that is eaten. Called the head, it is comprised of about eight to twelve segments, known as cloves. The papery skin is either white or flushed purple and the flesh white. If any part of the flesh is tinged green, it should be cut away because it will be bitter and may cause an upset stomach. The cloves should fill the skin; if they 'rattle', discard the head because this indicates that the cloves are dried and shrivelled. Do not refrigerate garlic; it should always be stored where air can circulate freely around it. The taste of garlic is strongest when eaten raw, with the flavour becoming more delicate the longer it is cooked. The flavour is also subtler if the clove is left intact instead of being sliced. Dried or pickled garlic is also available, although these are far inferior to the fresh kind. To eradicate the smell of garlic from your hands, rub them with a stainless-steel item.

'Garlic', coloured etching by Dubois (n.d.)

from Flore Médicale

C. L. F. PANCKOUCKE (1814)

CABBAGE FOR HEALTH

Traditional herbalists believed that the cabbage was beneficial to health, and it seems to have been something of a cure-all. People were advised to consume cabbage in a variety of ways, depending on the condition. Those bitten by an adder should have cabbage juice in wine. Immersed in honey it brings relief to a hoarse throat or lost voice. If the juice is boiled up with honey, the cooled liquid can be dropped into the corner of the eye to clear cloudy or dim vision. One of the least-pleasant sounding cabbage cures involves the mixing of the ashes of burned cabbage stalks with old pig's grease; it is said this concoction is 'very effectual to anoint the side of those that have had long pains therein, or any other place pained with melancholy and windy humours.'

ONION RHYME

This is every cook's opinion –
No savoury dish without an onion,
But lest your kissing should be spoiled
Your onions must be fully boiled.
JONATHAN SWIFT

CARROT CURES

Herbalists traditionally believed that the wild forms of plants were far more beneficial in medicinal preparations than their cultivated counterparts. Wild carrots in particular were thought helpful in alleviating flatulence, and urinary and menstruation problems, banishing kidney stones, as well being helpful for women who were having difficulty conceiving a child. Wild carrots are much smaller than their cultivated relatives, their texture is firmer, and their taste rather sharp.

Today, considerable medical and scientific research has been carried out on the cultivated carrot and it is considered to be a highly nutritious food. The French town of Vichy was famous during the 18th century for promoting the daily consumption of carrots as a cure for those with overloaded digestive systems. The dish called Vichy carrots still appears on menus today. Raw carrots, again eaten daily, are recommended for those with high blood cholesterol as well as for smokers, because they may cut the risk of lung cancer.

There remain a great number of old French varieties still available for the home gardener to grow, including 'Chantenay Red Core', 'De Colmar', 'Jaune de Doubs', and the very ancient 'Saint Valéry'.

CUCUMBER SUPERSTITIONS

Among the strange practices associated with the cucumber is the belief that the vigour of the growth of a cucumber vine is directly related to the strength of the sower. Consequently, it is said that only young and virile men should sow seeds of cucumbers. By contrast, menstruating women should neither look at, nor go near, the plants because this would cause them to wilt and wither instantly.

MARROW OR COURGETTE

Novice cooks are often a little confused about the distinction between these two vegetables. When does a courgette (called a zucchini in the United States) cease to be a courgette and become a marrow? Well, usually very quickly once overlooked by the gardener, left on the plant to grow and allowed to swell and expand into a huge watery Zeppelin-shaped vegetable! Tender and juicy courgette (*Cucurbita pepo*) come in a lovely array of colours including yellow as well as pale and dark green, and can be cooked or eaten raw in salads when young. If you wish to grow a colourful group of heirlooms why not try 'Black Beauty' (a very dark green Italian variety), 'Nimba' (a Polish courgette whose young fruits are light green with dots) and 'Yellow Crook Neck' (a yellow curved variety from the United States). Huge marrows have assumed something of a comic air and have long been entered in giant marrow contests at country fairs and shows. Their high water content makes them rather tasteless and they are normally stuffed with strong-flavoured ingredients before baking. In the past, seeds of some old varieties were sold as marrows, rather than as courgettes, such as the 'Boston Marrow Improved' and 'Mammoth'.

AMAZING CORN MAZE

In the 17th century the hedge maze became a popular pastime for the playful rich in Europe who loved the thrill of getting lost within the confines of its tall evergreen walls. In the 21st century, people are less patient and can't wait the ten or more years it takes for a yew hedge to become an impenetrable screen, so instead they plant corn mazes. It takes no more than a season for corn plants to grow as high as an elephant's eye (well, almost) and provide the perfect 'instant maze'.

COOPERATIVE PLANTS

Native Americans traditionally grew tall corn, which acted as a support for climbing beans such as butter, kidney, or pinto and these in turn were followed by squashes or pumpkins.

THE CHARD
Beta vulgaris subsp. *cicla*

CHENOPODIACEAE

CHARD IS A DELICIOUS, if somewhat neglected, vegetable that originates from the Mediterranean region and the Near East. It was widely cultivated by the ancient Greeks and Romans, was popular in France in the Middle Ages, then spread more widely throughout Europe in the 17th century. A close relative of the beet, chard is cultivated for its stalk and leaf instead of its root. It is also referred to as leaf beet, spinach beet, or silver beet and frequently chard is misleadingly identified as Swiss chard. It is not the same as perpetual spinach, which is similar in appearance but less heavily veined. It is sometimes used as a substitute for spinach, although many consider it to have a superior flavour, but, unlike spinach, it should always be cooked. Rich in vitamins A and C, chard is also an excellent source of iron, calcium, carotenoids, magnesium and potassium.

The most commonly commercially available chard is the white stalked cultivar, but beautiful red-, yellow-, pink- and orange-stalked cultivars are also grown, especially by gardeners who sometimes plant it in their borders as well as the vegetable patch! Among the heirloom cultivars still available is the silver-stalked 'Argentata', which is an Italian strain with a wonderful flavour. The rather confusingly named 'Rhubarb' (sometimes also called 'Ruby') is a firm favourite, due to its dramatic coloration and looks especially stunning when grown alongside the Danish cultivar 'Yellow Dura'.

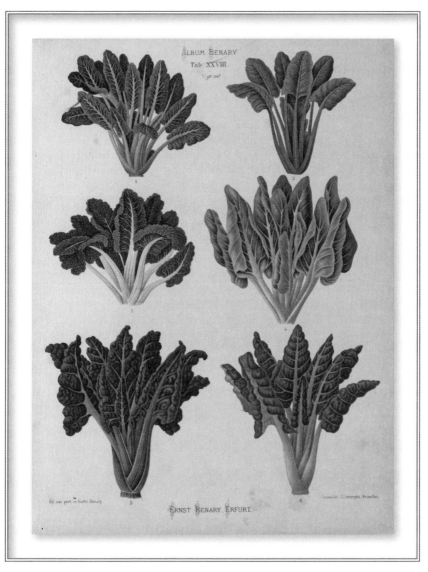

'Chard', by Ernst Benary (1819–1893)

from Album Benary

COLOUR LITHOGRAPH BY G. SEVEREYNS

SOME LETTUCE FACTS

Today, the lettuce is one of the two most popular vegetables consumed by Americans (the other is the potato). On average 14kg/30lb of lettuce are devoured per person each year. Romaine lettuce is the highest in vitamin A and C content while the butterheads score more for iron. By weight, 95 per cent of a lettuce's leaves consists of water, which accounts for their crispy and crunchy texture, as well as their propensity to wilt when not too fresh.

CELERY & CELERIAC

The original wild celery was a marsh plant native to Europe and used by the ancient Greeks and Romans for purifying the blood. Celery (*Apium graveolens*) was cultivated in Italy from the 16th century and has remained a very popular vegetable ever since. Green cultivars (known as Pascal) are slightly bitter tasting while the whiter cultivars (golden) tend to be crisper. The flavour of celery is much improved if harvested after the first frosts. An unusual heirloom variety is 'Giant Red', with stems that are tinged dark red, then turn pink when cooked.

Very similar in flavour to the long-stemmed celery is the root vegetable celeriac (*Apium graveolens* var. *rapaceum*) also called celery root, céleri rave, knob celery or German celery. With its knobbly and hairy turnip-shape, it is certainly less elegant than the long-stemmed celery but tastes wonderful in soups, stews or if puréed with butter.

SEA KALE

Sea kale (*Crambe maritima*) is a wild plant that grows along many of the shorelines of Europe. It is a brassica that has been cultivated in English gardens, from cuttings and seeds, since the early 18th century and it was mentioned in Thomas Jefferson's *Garden Book* of 1809. Sea kale was a popular food, often served at the tables of the rich, until the early 20th century but today it is no longer widely consumed. Its shoots are blanched prior to preparation, most usually in specially designed earthenware pots with removable lids. Manure was piled high around the pots to increase the heat, or sometimes they were grown on a hot bed or in a greenhouse.

Sea kale's taste is between that of asparagus and cauliflower. Cooked, the vegetable is frequently served with a simple dressing of lemon-flavoured melted butter or a hollandaise or a béchamel sauce. It should not be confused with chard (*Beta vulgaris* subsp. *cicla*), which is also sometimes referred to as sea kale beet.

THE TRUE YAM

The sweet potato (*Ipomoea batatas*) is often wrongly referred to as a yam. The true yam (*Dioscorea alata*) is a tropical rootcrop that grows as an herbaceous vine with an edible tuber below ground, also known as water, winged or purple yam. In New Guinea and Melanesia the vegetable plays a part in rituals and ceremonies with specially grown yams that can weigh over 55kg/120lb and reflect the status of the grower within the community. Likewise, it is the underground tuber of the air potato (*D. bulbifera*) that is edible. Native to Africa and Asia, they are used to treat eye infections.

THE VERTICALLY CHALLENGED CABBAGE

Brussels sprouts were introduced into North America in the early 19th century, where, due to their appearance, they were regarded as a kind of miniature cabbage. In an attempt to promote their popularity, a New York vegetable seller hit upon the bright idea of employing the services of a circus midget called Tom Thumb to advertise the new 'Tom Thumb Cabbages', an early example of celebrity endorsement.

FAST FENNEL

The Greek word for fennel is *marathon* and, following their victory over the Persians at the Battle of Marathon in 490 BCE, a Greek runner raced the 42km/26 miles to Athens to deliver the triumphant news. As the battle was fought in a field of fennel, the vegetable and the long-distance run became forever linked.

SOME SHALLOT FACTS

The shallot (*Allium ascalonicum*) was called the Ascalonian onion by both the Greeks and the Romans, named after Ascalon in Palestine. It is one of the tastiest members of the onion family, is elongated in shape, and comes in an array of colours, including grey, pink and golden brown. The last type keeps best, while the first does not store as well but has the richest flavour, being much preferred over other varieties by French chefs, for whom the shallot is a kitchen staple.

All shallots have a more mellow taste than their larger round relative. Unlike ordinary onions, shallots become bitter-tasting when fried; so they should always be stewed. One particularly elongated heirloom cultivar is variously known as the banana shallot in Britain and as frogs legs in the United States.

THE HORSERADISH
Armoracia rusticana

BRASSICACEAE

HORSERADISH ORIGINATED in southeastern Europe and western Asia but is now widely grown in many areas of the world. It was cultivated and used by the ancient Egyptians and Greeks and features in the latter's mythology. Legend tells that the Greek god Apollo was told by the Delphic oracle that what we think of today as the humble horseradish was worth its weight in gold. It is one of the 'five bitter herbs' that Jews were instructed to eat at Passover. In the Middle Ages it was used in Britain, Germany and Scandinavia both as a culinary accompaniment to meat and to disguise the flavour of tainted or spoiled meat. Horseradish was taken to the New World by the early colonists and in the 17th century brewers mixed it with the herbs wormwood and tansy to create a horseradish ale, which they dispensed to weary travellers.

Horseradish is a hardy perennial that can be difficult to eradicate once established, so be sure to plant it only where you want it to flourish for many years to come. It is the long tap root that is used in cooking. The root has black skin, but the flesh is white and discolours quickly once cut, unless plunged into vinegar. The pungent and peppery-tasting root is grated and then ground and mixed with oil, vinegar and cream to make a sauce that is eaten with meat or vegetables. It can be bought fresh or already ground.

'Horseradish', gouache by Jean-Charles Verbrugge (1756–1831)

from Collection du Régne Végétal

BARON JOSEPH VAN HUERNE (1790–1820)

CRESS & WATERCRESS

Cress (*Lepidium sativum*) is European in origin and has a long history, dating back at least to the Romans. Grown with mustard (*Sinapis alba*), it has become a garnish that is used to spice up a sandwich or simply adorn the side of a plate. Among the few old cultivars available is 'Persian Broadleaf Cress', which originates from Iran, where it is also called *shah hi* or 'royal food'.

Both wild and cultivated watercress (*Nasturtium officinale*) is available, grown in clear running water. One of the tastiest dark green leafy vegetables, watercress has a pungent, peppery flavour and is eaten raw or wilted. Its nutritional value is high, and it is rich in iron, calcium, sulphur, iodine, carotenoids and essential fatty acids. American or upland cress (*Barbarea praecox*), also called land cress, winter cress and Belle Isle cress, is a native of Europe and North America and often used as a substitute for watercress.

POTATO GUN

This may be an apocryphal tale, but the American murderer, gangster and bank robber John Dillinger (1903–34) is said to have carved a replica gun from a potato and used this as a prop to help him break out of jail. Some accounts even go so far as to name the cultivar of potato, the 'Russian Banana'!

TYPES OF GARLIC

Young garlic that is harvested in spring is known as 'wet garlic'. It resembles a small leek and has a much milder flavour at this stage than if left to develop. It can be eaten raw, like a scallion, but should be added later during cooking than the more mature types. The larger cultivar, known as elephant garlic, is also milder than the most commonly used standard size. This type of garlic is harvested in summer, then dried and stored. Single cloves are planted in winter (although traditionally on the shortest day, December 21). They then develop and swell to form firm bulbs.

SOME OKRA FACTS

Food writer Jane Grigson calls okra, also known as lady's fingers or gumbo, 'the most elegant of vegetables'. It consists of the long, green, immature, five-sided pods of the plant *Hibiscus esculentus* that are eaten. Originally a native of Africa, the plant was transported to North America with the slave trade where it became a staple of the cuisine of the Southern states. Okra also appears in Indian and Middle Eastern dishes. It is characterized by a rather sticky and syrupy texture when cooked and is useful for thickening soups and stews.

HOW TO SAVE YOUR OWN HEIRLOOM SEEDS

Method for saving the seeds of 'wet vegetables', such as tomatoes and squashes.

1. Open the ripe vegetable and carefully scrape out all the seeds and pulp.
2. Place in a jar and leave to ferment for several days, stirring occasionally.
3. Add some more water to the jar and discard any seeds that float (these will not be viable).
4. Drain the contents of the jar through a fine strainer.
5. Carefully wash away the pulp and drain the seeds.
6. Place the seeds on a ceramic plate, leave to air dry in a warm, but not hot, place, for a few days.

Method for saving the seeds of dry seed pods, such as beans and peas.

1. At the end of the growing season leave the pods to dry out on the plant.
2. If the weather turns wet or frosty, uproot the whole plant and hang in a cool, dry place.
3. Once thoroughly dry and brittle, cut the pods from the stems and carefully remove the seeds.
4. Sort through the seeds and save only those that are whole and undamaged, then gently shake them free of any dust or debris.

Method for saving seeds from flowering plants, such as lettuce.

1. Allow the plant to bolt, so that it forms flowering seed heads.
2. As the seeds begin to dry, inspect the plant regularly and shake any seed heads that are ready into a paper bag. Alternatively, cut the whole plant at the base, suspend over a container, and leave the seeds to ripen for a few days. They will then fall into the container.
3. Shake seeds through a fine strainer to remove dust and debris.

Whichever method is used, store the seeds in a cool, dry place in sealed jars or paper envelopes. Ensure they are well protected from mice and insects. Label with plant name, cultivar and date.

THE PUMPKIN & SQUASH
Cucurbita spp.

CUCURBITACEAE

THE LARGE GROUP of plants known as pumpkins and squashes originate from South and Central America, where they have made an important dietary contribution for thousands of years. They basically fall into two distinct groups: the winter types are hard shelled and have good storing properties, while the summer types have soft skins and must be eaten when fresh. Pumpkins are sometimes cultivated for their potentially vast competition-winning size and weight, but many think squashes have the superior taste.

Pumpkins and squashes come in a huge array of sizes, shapes and colours. They range in size from the large 'Mammoth Gold' to the tiny 'Jack Be Little'. Virtually all the colours of the rainbow are represented, including variations of orange, red, yellow, blue, green, white and pink, which may appear as stripes or in mottled and marbled effects. The texture of the skin or shell can be smooth, ribbed or warty, while the squash's form can be globe-shaped, oval, straight or crook necked, banana-shaped, scalloped edged or turban capped.

Pumpkins and squashes are used as ingredients in both sweet and savory dishes. Their nutritional value is good and they provide plenty of fibre, although vitamin content varies substantially from type to type. For instance, the popular butternut squash (which falls into the *Cucurbita moschata* group) contains 80 per cent more vitamin C than an acorn squash (C. *pepo*).

Citrouille moyenne nommée le Pâté, ou Bonnet carré

'Pumpkin', gouache by unknown Flemish artist (n.d.)

from Collection du Régne Végétal

BARON JOSEPH VAN HUERNE (1790–1820)

OTHER ONIONS

Most cooks are all too familiar with the large yellow- and red-skin types of onions and the scallion, or green onion, with its mild-tasting small bulb, but there are others. The Welsh onion (*Allium fistulosum*), despite its name, hails from Siberia. It is popular in Asia, where it is also known as the Japanese leek. You may also hear it referred to as the everlasting onion, because the plant produces a crop for several seasons. It produces a tight bunch of long thin bulbs, used in salads much like scallions. The potato onion (*A. cepa aggregatum*) forms clusters of bulbs that are larger than a shallot, just under the surface of the soil. As the clusters rise to the surface, the soil should be gently brushed away so they can ripen in the sun, but be careful not to uncover the roots. Potato onions have a mild flavour and good storing properties.

Perhaps the most curious of all is the Egyptian walking onion (*A. cepa proliferum*), also known as the tree onion, or sometimes the topset onion. A native of Canada, it can withstand very harsh winters. The 'trunk' of the 'tree' is actually a hollow stalk that grows to around 1.2m/4ft in height and the small, strongly flavoured onion bulbs form on this.

SAMPHIRE

This European vegetable is gradually appearing with increasing frequency on restaurant menus and in stores, especially fish dealers. Rock samphire (*Crithmum maritimum*) grows wild on seaside cliffs and rocky shores, and is cooked or pickled, while the unrelated marsh samphire (*Salicornia europaea*), found on salt marshes, is either cooked or used in salads. Its name is derived from the French *herbe de Saint-Pierre*, St. Peter's herb. Another name for marsh samphire is glasswort because it was used in glass making and it is also sometimes referred to as sea beans or baby asparagus.

J^U MP^I NG BEA^N S

The vegetable oddity known as Mexican jumping beans are not really beans at all, nor do they jump! In fact, they are the seed pods of the mountain shrub *Sebastiania pavoniana*, into which the tiny larva of a moth has bored a hole, then made its home within. It is the jerking of this larva inside that causes the characteristic twitching and rolling that so resembles jumping and jiving.

HOW TO GROW HEIRLOOM PUMPKINS & SQUASHES

Perhaps more than any other vegetable, the sheer variety in shape, size, colour and form of pumpkins and squashes make this group of plants a 'must grow' for all gardeners lucky enough to have the space. Even if one chooses to grow a range of different cultivars each year, it is unlikely one will ever exhaust the tempting list of available seeds. Apart from the taste, culinary versatility, and long-term storing properties of winter pumpkins and squashes, many are so beautiful to look at that they are also worthy of display as indoor decoration.

1. Sow seeds in pots in early spring indoors, or after all danger of frost has passed if sowing directly into the ground.
2. The 'hill planting' technique has traditionally been used for growing pumpkins and squashes. Once the ground has been prepared (thoroughly weeded, with plenty of compost dug in), mound the soil to create a round bed approximately 60cm/24in wide and 15cm/6in high. These are sprawling plants, so leave about 3m/10ft between the hills.
3. Plant one pot-grown plant firmly on each hill. If sowing directly, sow four seeds per hill at a depth of 2.5cm/1in.
4. As the seedlings develop, thin out the weakest, leaving the two most vigorous to grow on.
5. Water well and regularly but do not drench the small plants, mulch with more compost, then stand back and watch as the plants begin to romp away.
6. As the pumpkin or squash develops, place a large tile or a bed of straw underneath each vegetable to protect them from the damp ground and prevent rotting.
7. Once the fruits are fully developed and the skins begins to harden, cut them off the main plant, ensuring that a stem of about 7.5cm/3in long is left attached, and put in a dry place to cure in the sun for a few weeks. This will significantly improve their storing properties.

THE CHICORY

Cichorium intybus

ASTERACEAE

AS COOKS AND CHEFS become ever more adventurous in their quest to create innovative and colourful salads, chicory is once again becoming a popular leaf vegetable. Chicories fall into three distinct groups, although their names are sometimes confused because the names differ, depending on where you live. The pale Belgian chicory must be blanched to reduce its bitterness prior to eating. 'Witloof' is the variety that is usually grown (indeed chicory is often referred to as 'Witloof' in America and Australia) and the blanched heads are called chicons. The French refer to these types as endive. Green-leaved chicories, which resemble a Cos or Romaine lettuce, need no blanching, although to avoid bitterness they should be cut before they grow too large. Typical varieties are 'Spadona' and 'Sugarloaf', both originating from Italy. The glamorous red types have red leaves with contrasting white spines ('Red Treviso' is one of the oldest varieties, and again hails from Italy) or they are mottled in shades of green, red and white (such as 'Castelfranco', an 18th-century Italian variety). These also go by the names of red chicory or radicchio.

The roots of chicory were baked and ground then used as a substitute for expensive coffee beans in Europe during the 1820s and are still used as an additive to coffee today. Some brewers also used roasted chicory, adding it to their stout beers for extra flavour.

'Chicory', colour lithograph by Elisa Champin (n.d.)
from Album Vilmorin
VILMORIN-ANDRIEUX & CIE (1850–1895)

CABBAGES FROM JERSEY

The Jersey cabbage is a real vegetable curiosity and is sometimes referred to as Jersey kale or cow cabbage. It has a single stalk that can grow as high as 3m/18ft and sprouts cabbage-like foliage at the top. Sheep who were fed the leaves were said to produce particularly fine and silken wool. The stalks were cut, dried, then used as walking sticks by the island's inhabitants.

BEANS & HERBS

Just as basil is commonly teamed with tomatoes, the herb summer savory (*Satureja hortensis*) is an ideal companion in all kinds of bean dishes. In German markets it is not unusual to see bunches of beans tied together with sprigs of the herb for sale. Indeed the German for summer savory is *bohnenkraut*, literally 'bean leaf'.

HAMBURG PARSLEY

Unlike the more common types of herb parsleys, which are grown for their leaves, Hamburg parsley (*Petroselinum crispum*) is cultivated for its roots. An alternative name is turnip-rooted parsley or simply parsley root. Shaped like a somewhat thin parsnip, with grey colouration, its delicate taste is reminiscent of celery and parsley leaves with a hint of nuts. Hamburg parsley should be thoroughly washed, but not peeled, before cooking. It puts in an appearance in traditional recipes from several European countries including Bulgaria, Germany, Poland and Russia.

A BRIEF HISTORY OF ORACHE

Thought to originate from Central Asia and Siberia and dating from prehistoric times, orache (*Atriplex hortensis*) deserves a place in many more kitchen gardens than it currently occupies. Although popular in Europe during the Middle Ages, orache had largely fallen from favour by the 18th century. However, by the following century it was widely grown in North America mainly due to Fearing Burr's *The Field and Garden Vegetables of America* of 1863. In this volume Burr listed sixteen varieties of orache with leaf colours ranging from green and yellow to red. Indeed, it is such an attractive plant that it is often planted as an ornamental. It is hardy and, as its common name, mountain spinach, hints, is a leafy vegetable much like spinach (although less troublesome to grow and with a sweeter taste).

Over time orache has gathered a selection of intriguing names that bear little relation to each other, among them the good lady of the garden, the illustrious, the prudish woman and the cabbage of love!

A BRIEF HISTORY OF MUSHROOMS

Strictly speaking, a mushroom is not a vegetable but the fruitbody of a fungus that grows in the earth. Mushrooms appear in Egyptian hieroglyphics and were considered as food fit for pharaohs. The ancient Romans likewise considered the mushroom to be a great delicacy.

Only wild mushrooms were consumed until the 17th century, when a French botanist demonstrated that the mysterious fruitbodies grew from spawn, after which mushroom cultivation became something of a French speciality. Cultivars were introduced into North America during the late 19th century.

There are many types of mushroom with colours that range from brown to cream, grey and white. One of the most popular mushrooms is the small button type, while the less familiar one known as the lobster is said to have a slightly fishy taste! Shiitake mushrooms have become widely available in the West in recent years and are highly nutritious. In Asia they have long been known as the 'elixir of life'.

All mushrooms contain calcium, iron, protein, phosphorus and potassium as well as the B vitamins. As they are composed of about 80 per cent water, they readily absorb the flavours of other ingredients in cooking.

BEAN SUPERSTITIONS

In medieval times a bean would be hidden in a cake at Twelfth Night celebrations. The lucky man who found this bean in his slice of cake was then anointed 'Bean King' for the following year and was granted all kinds of favours. Gardeners and farmers alike were advised to line a bean trench with horsehair prior to sowing, thus ensuring an excellent crop. Several sayings have acted as useful *aide mémoires* at bean sowing time, such as the warning 'Sow beans in the mud, they'll grow like wood'. Another oft-cited phrase is 'Sow one to rot, and one to grow, one for the pigeon, and one for the crow'.

EDIBLE PICTURES

Giuseppe Arcimboldo (1527–93) was a Milanese painter who produced fantastic portraits composed of fragments of landscape, flowers, foliage, fruits and vegetables. Under his expert hand, squashes became elaborate coiffed hair, a row of parsnips formed a beard, while onions represented plump cheeks. These proved very popular in his day and were later applauded by the Surrealists. Perhaps most surprising of all is that his patron, Rudolf II of Austria, was happy to have his own 'portrait' created in this way!

THE BEET

Beta vulgaris

CHENOPODIACEAE

ALTHOUGH THEIR PLACE and date of origin is somewhat hazy (some think they came from the Nile and Indus region 2000 BCE while others claim northern Europe), our modern-day beetroot most probably derives from *Beta vulgaris maritima*. This wild beet grew in Mediterranean coastal regions and also gave rise to chard, a close relative of beetroot. The Greeks ate the leaves (which are as nutritious and tasty as the roots). The Romans cultivated the enlarged taproot and it was this that was eaten as food, rather than the leaves, and also used for medicinal purposes. This Roman connection may be why it was called 'Romaine beete' in Tudor England.

It is the red beetroots that are most commonly available but home growers can easily obtain seed of orange, yellow and white varieties, although the taste differs little between them. The variety known variously as Egyptian Turnip Rooted or Egyptian Flat is a much flatter shape than the usual beetroots. Its roots grow just underneath the soil with a very dark violet skin and deep purple flesh. They have an excellent taste. Beetroots are rich in iron, calcium, carotene, magnesium, phosphorus, vitamin A and the B complex vitamins. The mangle, also known as mangold or mangel-wurzel, is a chance hybrid of beetroot and chard that originated in Rhineland Germany and is mainly used for cattle feed.

'Salad beets or beetroot', by Ernst Benary (1819–1893)

from Album Benary

COLOUR LITHOGRAPH BY G. SEVEREYNS

THE PARSNIP

Pastinaca sativa

APIACEAE

PREHISTORIC REMAINS of the wild parsnip have been found at archeological sites in Germany and Switzerland. The vegetable is thought to have originated in the eastern Mediterranean regions and is now widely available in many countries. Precise identification of its origin and its subsequent spread and culinary use is somewhat obscure, because many texts appear to confuse the parsnip with the carrot! The former was introduced into North America in the early 17th century but sadly it has yet to gain the same popularity there that it enjoys elsewhere, for instance in much of northern Europe. The French and British in particular value this sweet-tasting root. In Russia the parsnip is known as a *pasternak*, coincidentally sharing its name with one of the country's most revered authors (Boris Pasternak 1890–1960).

The parsnip has creamy white skin and flesh and is shaped like the common carrot with a long, tapering root, but it is usually much thicker around the shoulders. It has a sweet, often slightly nutty, succulent taste. A very hardy vegetable, parsnip can be left in the ground throughout the colder months, which improves its flavour. Those tempted to grow heirloom parsnips should perhaps try to track down the seeds of a cultivar called 'The Student'. Rather intriguingly, this is said to be the sweetest parsnip of all, with a distinct and surprising aroma of lavender flowers!

'Parsnip', colour lithograph by Elisa Champin (n.d.)

from Album Vilmorin

VILMORIN-ANDRIEUX & CIE (1850–1895)

ONION TEARS

৯৫

It is one of nature's many ironies
that one of the most commonly used vegetable
ingredients regularly reduces the cook to tears. The very
first step in so many recipes is to 'finely chop the onion', after
which subsequent instructions become a blur as the eyes fill with
water and tears begin to flow! When an onion is sliced, the individual cells
of the onion flesh are broken, causing certain enzymes to break down and
amino acid sulphoxides to form unstable sulphenic acids. Due to their
instability, these acids quickly re-form, in part as a volatile gas that is released
into the air. This quickly reaches the eyes, causing irritation and stinging,
followed by floods of tears. Preventive measures include chilling the
onion prior to cooking (this helps to limit the strength of the
reaction) or cutting the onion in a bowl of water or under the
faucet (so keeping the gas in the water instead of
releasing it into the surrounding air), although
this has the added danger of

৯৫

COTTAGER'S KALE

Here is an intriguing description of a delicious-sounding vegetable, cottager's kale, that
appeared in the 19th-century *Beeton's Gardening Book*. Fortunately, it is still available from
heritage and heirloom seed suppliers. 'This is a variety of the tall cavalier cabbage, which
was raised at Sherburn Castle, Oxfordshire, from Brussels sprouts. Crossed with one of the
varieties of kale, it was submitted to the Horticultural Society in the spring of 1858, and is
said to be the most tender of all the greens, and of exquisite flavour. It stands four feet high
when full-grown, and should be allowed an equal space to grow in, being clothed to the
ground with immense rosette-like shoots of a bluish-green tint, which, when boiled,
become a delicate green. The seed should be sown in early spring, and the plants should
have a rich deep soil assigned to them'.

A THING OF BEAUTY

In 1947 Marilyn Monroe, who was then only a Hollywood starlet, was crowned the first
'Queen of the Artichoke' in a contest organized by the town of Castroville, California,
which claimed itself to be the 'Artichoke Centre of the World'.

PRESERVED VEGETABLES

However superior the taste and texture of recently harvested produce, it is inevitable that the busy cook must sometimes reach into the cupboard for a can of sweetcorn, or that handy bag of frozen peas. The two inventors to be thanked for such miracles are Frenchman Nicolas Appert (*c.* 1750–1841) and American Clarence Birdseye (1886–1956). Nicolas Appert was a chef, confectioner and distiller who successfully experimented with preserving foodstuffs in hermetically sealed containers, such as bottles sealed with corks and wax. Appert's discoveries were taken a stage further by a British merchant, Peter Durand, who, in 1810, patented his own version: the tin can.

Clarence Birdseye's place in culinary history was assured after a fishing trip with the Inuit of Labrador, Canada, during which he observed how his catch froze almost instantly, yet, once defrosted, tasted just as good as fresh fish. Once home, after further experimentation, Birdseye set up the General Seafood Corporation. His determination to retain the flavour, colour and texture of preserved food revolutionized both the nutritional quality and availability of what we eat today.

SOME TURNIP FACTS

Turnips (*Brassica rapa* var. *rapa*) vary greatly in shape, size and colour, with round, cylindrical, yellow, and white cultivars available. Although they are grown for their roots, their green-leaved tops can also be used as a leaf vegetable. The young turnips of spring are far superior to the old. The navette is a French turnip, which is long and thin, more like a carrot, and is often seen bunched and tied together in French markets. According to legend, the Halloween jack-o'-lantern was originally carved from a turnip, rather than the more colourful pumpkin.

VEGETABLE SQUASH

The curious spaghetti squash (*Cucurbita pepo*), also known as noodle squash, spaghetti marrow, vegetable spaghetti, gold string melon, fishfin melon, as well as the hybrid term squaghetti, is a firm favourite with home growers who seek a bit of novelty in their crops. It has bright yellow or creamy white skin and, when ripe, can be boiled for 40 minutes, halved, then the spaghetti-like flesh scooped out and topped with a pasta sauce. The Chinese consider that its cooked flesh resembles shark's fin.

THE ASPARAGUS

Asparagus officinalis

LILIACEAE

The original wild asparagus hails from the fertile areas around the Nile and the Indus rivers. Later the Romans introduced it into Europe. In the Middle Ages it was cultivated in monastic gardens for both medicinal and culinary uses. While the ancient Greeks had used it to treat toothache and bee stings, the monks administered it for heart palpitations and as a diuretic. For the market gardeners of the Italian Veneto in the 16th and 17th centuries, asparagus proved to be an extremely profitable crop and it contributed considerably to the wealth of the region. During the same period London street traders sold bunches of spears under the common name of sparrow-grass. Early settlers took asparagus to the New World, where it only really gained wide acceptance in the early 20th century when the canning process made it available to the masses (although in quality terms, this is far inferior to freshly harvested spears).

Asparagus has always been regarded as a luxury food and a great delicacy. It is either white (as favoured by the Spanish) or dark green. Many types derived from old cultivars are still grown. The 16th-century 'Violet Dutch' from Holland, the 17th-century 'White German', with white tips, and 'Connover's Colossal' (from the 19th century) are all famous names in the annals of asparagus growing. Asparagus connoisseurs claim that the male plants produce the finest spears; 'Lucullus', 'Franklin' and 'Saxon' are all modern male cultivars.

Asparagus officinalis Lin.

'Asparagus', colour lithograph by Elisa Champin (n.d.)

from Album Vilmorin

VILMORIN-ANDRIEUX & CIE (1850–1895)

SOME KOHLRABI FACTS

Kohlrabi (*Brassica oleracea*) is either green or purple. Its very descriptive name is German and means 'cabbage-turnip'. It is cooked in the same way as turnips and has a similar, although many think more delicate, taste. Kohlrabi can also be eaten raw like an apple (or grated into salads) and should only be used when about the same size as that fruit. The young leaves can be used in a similar way to spinach. For some time kohlrabi has been particularly valued for its medicinal properties by the Chinese, who consider it a good tonic and beneficial to the body's equilibrium.

SOME LEGUME FACTS

Legumes (peas and beans) are some of the oldest crops grown on the Earth. Indeed 6,000-year-old beans have been discovered in Mexican caves, while broad beans were grown over 4,000 years ago in Mesopotamia and may even be the first vegetable to be cultivated by humankind. The ancient Greeks deemed beans to be a fit offering to their god Apollo and they even featured in the political process. When votes were cast, a black bean denoted 'for' while a white bean was a vote 'against'. Several dynasties in ancient Rome even took their names from legumes, including the families Cicero (chickpea), Fabius (fava), Lentullus (lentil) and Piso (pea).

Today beans are recognized as a great source of protein and feature strongly in vegetarian diets. They contain little fat and no cholesterol, and are plentiful in iron, calcium, phosphorus and fibre. Their medical qualities have long been valued for a whole array of conditions. Traditional herbalists considered 'garden and field beans' a valuable aid when treating acne, wrinkles, boils, bruises, kidney stones, inflamed wounds, sciatica and gout, while peas were thought 'to sweeten the blood'.

GARLIC MEDICINE

The medicinal qualities of garlic are legendary and were cited by ancient practitioners in China, Babylonia, Egypt, Greece and across the Roman Empire. In 1858 French chemist and microbiologist Louis Pasteur (1822–95) conducted experiments proving the antibacterial qualities of garlic. He discovered that about a quarter of a teaspoon of raw garlic juice killed bacteria (effectively as 60mg of penicillin). This knowledge was useful in the Second World War when, due to a shortage of penicillin, diluted garlic solutions were used to disinfect the open wounds of British and Russian soldiers, thus helping to prevent gangrene. Modern-day herbalists still prescribe it for a range of conditions including cancer, heart disease, high cholesterol, digestive problems, fevers and colds. However, the English tradition of placing garlic cloves in children's socks to cure whooping cough is still to be substantiated.

HOW TO MAKE AN ASPARAGUS BED

An asparagus bed is a wondrous thing. Once successfully established, a productive and well-tended bed will produce an impressive annual crop for up to thirty years. It is a slow process but it pays dividends in the long term, as asparagus is one of those vegetables that must be eaten as quickly as possible after harvesting. The proficient gardener-cook will set the pan of water to boil before going out to their asparagus bed to pull some stalks for supper.

1. First choose a site that can be left undisturbed for many years and allow about 1m²/1sq yd for each root.
2. Any well-drained soil will suffice, but the position should be exposed to full sun and be protected from harsh winds.
3. Dig over to remove all weeds and add in plenty of well-rotted manure.
4. In spring, after all danger of frost, dig out the planting trenches. These should be 25cm/10in deep by 25cm/10in wide. Position them about 1m/3ft apart and add more manure to the base of each trench. Top this with a thin layer of soil in a slightly raised mound.
5. Lay the asparagus root on top of the mound, being careful not to damage the root system. Cover the root with about 5cm/2in of soil but keep the green top of the bulb exposed to the light. Water well.
6. As the plants grow, gradually backfill the trench with soil, ensuring that any new shoots and leaves are above the soil level.
7. Once the trench is full, mulch well with garden compost and keep well weeded.
8. Do not harvest any spears that appear during the first two years (however tempting!).
9. In the autumn cut down the ferny foliage once it has died back.
10. In the third and subsequent years, harvest the spears just above ground level when they have reached about 15cm/6in tall. Do not cut the spears – just break them off – because cutting can damage other newly forming spears.

THE KOHLRABI

Brassica oleracea var. *gongylodes*

B R A S S I C A C E A E

THE KOHLRABI GROWS as happily in cool alpine valleys as in hot semi-desert regions. Its origins are uncertain but the ancient Roman author Pliny the Elder described a plant that resembles our familiar kohlrabi, which he called the Corinthian turnip. Kohlrabi has been cultivated and eaten since the 15th century in Europe but, although widely available today, it has never really enjoyed the popularity of its close relatives: Brussels sprouts, broccoli or cauliflower. The Chinese value its medicinal qualities.

Once harvested, kohlrabi can easily be mistaken for a root vegetable but, in fact, both the swollen stem and the leaves grow above the ground (both are edible). Several cultivars have been developed and they vary considerably in skin colour. 'Azur Star' is among the blue strains, 'Dyna', 'Purple Delicacy' and 'Purple Vienna' are violet-hued, while the white types include 'Rasko' and 'Trero'. There is also a white- to light green-coloured cultivar called the 'Superschmelz'. This grows to a considerable size yet still retains its sweet taste and tender flesh, whereas most kohlrabi should be consumed when they reach about the size of a golf ball – certainly no larger than a tennis ball – otherwise they tend to become woody in texture. If small enough, they do not need to be peeled and can be eaten raw, thinly sliced in salads, or lightly steamed and eaten with hollandaise sauce. Kohlrabi is a good source of potassium and vitamin C.

'Kohlrabi', by Ernst Benary (1819–1893)
from Album Benary
COLOUR LITHOGRAPH BY G. SEVEREYNS

ROCKET

Rocket is one of those old varieties of vegetable that have enjoyed something of a revival in recent years and now appears regularly on menus and at the vegetable counter. As recently as 1978 Jane Grigson wrote in her wonderful *Vegetable Book*, 'Rocket as an unremarkable item of salads is now eaten only in Mediterranean countries. John Evelyn grew it once in his kitchen garden, along with corn salad, clary, purslane, and all the other greens we have sadly allowed to disappear from our salad bowls.' There are two types, cultivated rocket (*Rucola coltivata*) and the wild variety (*Diplotaxis tenuifolia*). Its main distinction over other green leaves is its pungent, tangy and peppery taste (although frequent watering stops the leaves becoming too peppery). The wild variety is thought to have the superior taste while the more mature the leaf the stronger the flavour.

GROWING THE BIODYNAMIC WAY

Biodynamic gardeners grow their plants by following the phases of the Moon. Produce grown this way is said to be especially rich in nutrients as well as having better storing qualities. The varying types of crop are sown on particular days and are divided into groups according to the four elements. Roots (carrots, potatoes, radishes and so on) should be sown on Earth days, leaf plants (lettuce, cabbages and other leafy crops) on Water days, flowering plants (broccoli, globe artichokes, cauliflowers) on Air days, and 'fruit' (tomatoes, beans, peas) on Fire days. Fruit and vegetables harvested at the time of the New Moon are said to store well while those intended for eating fresh are better picked at Full Moon.

DRYING TOMATOES

Homegrown tomatoes taste so good, but unfortunately in many cooler regions the cropping season is relatively short. However, those lucky enough to have a bumper crop can preserve that wonderful flavour of summer by drying some of their tomatoes. All types can be used, although many claim plum tomatoes are best for this purpose. You need to select tomatoes of a similar size, to ensure that they dry out at the same rate. Remove all stalks, cut the tomatoes in half and place on a wire rack, not touching, set over a baking dish. Sprinkle lightly with sea salt and herbs (basil and marjoram team particularly well with tomatoes). Place them in the oven on the lowest possible setting for several hours. Check regularly, and remove from the oven when they feel dry to the touch but not brittle. Let them cool completely, then store in airtight containers.

TASTY TOMATOES

All those who grow their own tomatoes can testify to the incomparable pleasure of consuming a warm, ripe fruit straight from the vine. Such is the discrepancy between this tasty delight and its store-bought counterpart that it can sometimes be difficult to believe they are actually the same plant! Too often commercially produced tomatoes are picked unripe, then treated with ethylene gas to make them red although they are not ripe. However, they can be encouraged into ripeness by placing them upside down on a sunny windowsill. Tomatoes destined for canning are picked at the height of ripeness, ensuring both plentiful taste and nutritious quality.

BEAN SUPERSTITIONS

The bean plant may seem innocent enough entwined around a framework of canes, but this humble legume has long been associated with death and evil doings. Following a funeral, the ancient Greeks would sit down to a bean feast, and the vegetable also played a vital role during the exorcism of ghosts from dwellings. It is said that Pythagoras even believed that human souls became beans after death!

The Romans too thought the bean was an inauspicious vegetable. English coal miners believed that accidents were more likely to happen below ground when bean plants were in flower (because the souls of the dead were thought to live in bean flowers.) The flower scent was believed to induce nightmares and madness, and in Scottish legend, witches ride on beanstalks instead of broomsticks.

POISONOUS POTATOES

The nightshade or Solanaceae family of plants is huge, containing over 2,800 species. Non-culinary relatives include the tobacco plant and the prolific flowering climber known as morning glory. Among the edible vegetable groups are potatoes, tomatoes, aubergines and both sweet and chilli peppers.

However, be particularly careful as to which parts of these food plants are consumed and in what condition. The stems and leaves of potatoes and tomatoes are toxic while any part of a potato that is green must be discarded, because this discolouration indicates the presence of the glycoalkaloid poison. The chemical in chillis responsible for their hot and spicy characteristics is capsaicin, which is also a very efficient paint stripper! Be warned, because it can kill if eaten in sufficiently large quantities. More moderate consumers of hot and spicy foods can take comfort in the fact that chillis are also rich in vitamin C.

THE POTATO

Solanum tuberosum

SOLANACEAE

THERE HAS BEEN much debate and discussion concerning the exact origins and ancestry of the modern potato. Emerging consensus suggests the first edible potato grew high in the Andes and was consumed circa 5000 BCE. The crop was particularly suited to the hot days and cold nights of the high regions of South America. When Spanish raiders arrived in the region in the 16th century, the domesticated potato was one of the treasures they plundered and took back home, with its use spreading throughout Europe and beyond. Later the potato made the reverse crossing of the Atlantic to New England aboard the *Mayflower*.

Once available in Europe, potatoes were far from universally welcomed and only slowly gained wide acceptance. Their high toxicity if not stored and cooked properly led many to think that potatoes were poisonous. (They are, after all, members of the deadly nightshade family.) However, once it was understood that green potatoes should never be consumed, potatoes became a firm staple of many cuisines. Although potatoes can be cooked and prepared in numerous ways, their ubiquity and comparative cheapness has sadly resulted in their widespread abuse in many a kitchen, where they are either immersed in a deep-fat fryer as chips or fries, or transformed into processed mashed potatoes. With the resurgence of interest in the multicoloured heirloom cultivars, chefs and cooks are reclaiming the potato and it is again receiving the respect it deserves.

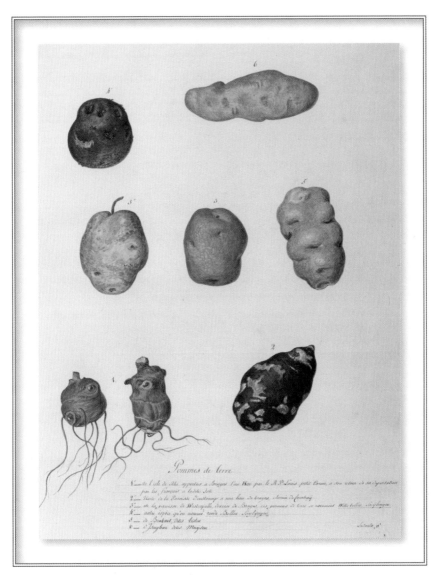

'Potatoes', gouache by unknown Flemish artist (n.d.)

from Collection du Règne Végétal

BARON JOSEPH VAN HUERNE (1790–1820)

GARLIC SUPERSTITIONS

Legends and superstitions surrounding the fragrant garlic occur in many countries and cultures. Perhaps their origin can be found in its Sanskrit name, which literally means the 'slayer of monsters'. In Transylvania it is reputed to ward off vampires, devils and werewolves (as well as amorous advances). It is used as a deterrent against the evil eye and witchcraft in China, Japan, India and Asia Minor. The Scots hang it in their houses on All Hallow's Eve (31st October) also to keep away evil spirits. When descending underground 16th-century German miners would take garlic cloves to protect them from demons, and in Korea travellers are advised to eat pickled garlic before traversing mountain paths, because it is said to deter tigers.

SUMMER SQUASHES

Unlike the winter squashes and pumpkins that have hard skins and keep for several months, summer squashes are soft skinned and are best harvested young and consumed quickly. They fall into three types: neck types (straight and crook-neck), Italian types (green and golden courgette and *cocozelle*), and the pretty scallop ones, which are also known as custard marrow. The many cultivars are variously bright orange, pale green, white with green-and-yellow stripes, and pure white. Pick them small and prepare like courgettes.

PUMPKIN SEEDS

The seeds found at the heart of pumpkins and squashes should not be discarded. Native Americans have long valued the seeds for their medicinal qualities – they are rich in magnesium, iron and zinc. Modern research has also revealed that the pulp contains anti-cancer carotenoids. The seeds can be dried and used as a highly nutritious and tasty addition to salads and breakfast cereals. To make your own, carefully wash and dry the seeds, removing all traces of the sticky pulp, spread on a baking tray, scatter with sea salt and roast in a hot oven for about five minutes. Allow to cool, then store in an airtight container.

HOW TO GROW HEIRLOOM POTATOES

Too many supermarkets offer a disappointingly small range of potato varieties Perhaps it is this that has led some cooks to treat the vegetable with a lack of respect, doing nothing more adventurous with it than boiling in salted water or deep-frying in fat! This is a great shame because, as the heirloom grower can testify, there is, in fact, a wonderful choice of exciting potato cultivars to explore, all with an array of different textures and colours.

1. To keep disease to a minimum, always grow potatoes using the crop rotation method (see page 14). They should never follow on from tomatoes, peppers or aubergines.

2. Prior to planting, potato tubers should be 'chitted' or sprouted. Lay the seed potatoes on trays indoors in the light. Soon nodules, or chits, appear. Once two or three have developed the tuber is ready for planting.

3. Early varieties should be planted in a hole at a depth of 10cm/4in, maincrops at 20cm/8in, with the chits uppermost. Traditionally wet newspaper or straw, along with some well-rotted manure, is added to the base of the hole before planting. Infill with soil, leaving a slight mound to help you identify the planting position. Plant at a distance of about a 30cm/12in apart.

4. As shoots emerge, ensure that the soil around each plant is gently earthed up, creating a ridge along the line of potatoes. This will prevent any light reaching the tubers growing below the soil.

5. For bumper yields, remove the flowers as they appear on the plants.

6. Once harvested early and second early varieties of potatoes do not keep for more than a couple of weeks, so lift these as needed. Always be careful not to spear the flesh when lifting potatoes with a fork.

7. Maincrop potatoes should be left to dry in the sun for a day or so after lifting.

8. Store in paper or hessian sacks in a cool, dry place, ensuring they are not exposed to light.

THE CABBAGE

Brassica oleracea var. *capitata*

B R A S S I C A C E A E

THE ORIGINAL WILD cabbage is a native of the coastlines of northern Europe. It has long been a staple of diets there and was certainly consumed by the occupying Romans. Over the centuries the cabbage was widely and enthusiastically domesticated and there are now a whole range of forms and cultivars. Cabbage is categorized both by its season, either spring, summer, autumn or winter, and by their form and leaf type, whether semi-hearted, green-hearted, hard white or red.

Savoy cabbages are among the loveliest of all varieties, with their closed hearts enwrapped in large, curly leaves with attractive serrated edges. They can be eaten raw or cooked and their generous leaves are ideal for stuffing. Spring greens are far smaller, loose leafed and form no hearts. The hard white types, also known as Dutch Whites, are most commonly used for making coleslaw and sauerkraut. Red cabbage looks as good growing in the garden as it does on the plate. The smooth, shiny red leaves are curled tightly around the heart and have contrasting white stems which make intricate patterns once cut. In Britain it is traditional to pickle red cabbage but it also makes a delicious vegetable accompaniment when baked slowly with chopped apple and seasoned with spices. Like many of the brassica family, cabbages offer exceptionally high levels of nutrients and are thought to have properties that inhibit cancer.

Brassica oleracea

'Early Savoy cabbage', colour lithograph by Elisa Champin (n.d.)

from Album Vilmorin

VILMORIN-ANDRIEUX & CIE (1850–1895)

FREUDIAN VEGETABLES

Sigmund Freud (1856–1939) propounded the theory that to dream of eating was symbolic of the sexual act. In cheap popular culture many vegetables provide a visual shorthand for phallic symbols, and several superstitions claim that phallus-shaped vegetables are, in fact, aphrodisiacs. In the esoteric world of dream interpretation, particular vegetables are said to denote a whole range of meanings, although their origins remain nothing if not obscure. For instance, should you encounter a cauliflower in the Land of Nod, expect an act of generosity; a green pepper signifies a cool thinker, while a pumpkin represents playfulness.

SWISS CHARD & ITS BEET BROTHER

In botanical terms beetroot (*Beta vulgaris*) and chard (*Beta vulgaris* subsp. *cicla*) are the same plant. Although their appearance and taste is very different they both come in a wonderful array of colours. Commercially produced beetroot is commonly a rich deep-red colour but home growers can opt for exciting varieties such as the white-fleshed 'Snow White' (also known as 'Albina Vereduna'). The 19th-century 'Golden' lives up to its name, having vivid orange roots that turn a golden yellow when cooked. Today packets of the mixed seeds of chard are often sold as 'rainbow' with the rich green leaves born on white, red, yellow, pink or orange stems.

SHAKER SEEDS

The 19th-century dissident religious group the United Society of Believers in the First and Second Appearance of Christ (more commonly known as the Shakers) were the first commercial producers of plant seeds in North America. Renowned for the quality of their seeds, quantities were sold in bulk to farmers in large cloth sacks, and the domestic kitchen gardener could buy small amounts packed in paper envelopes, the forerunners of today's ubiquitous seed packets.

In 1835 the United Society published *The Gardener's Manual*. Both a seed catalogue and an instruction booklet, it was marketed through their seed distributors and sold 16,000 copies, with a revised edition appearing in 1843. The business was extremely lucrative, with the New Lebanon community in Columbia County, New York, making a staggering $10,000 in a single year. Among the seed cultivars offered were the pole bean 'Clapboard', the carrot 'Altringham', the squash 'Winter Crookneck', and the turnip 'Long Tankard'.

A BRIEF HISTORY OF THE SWEET POTATO

Despite its name, the sweet potato (*Ipomoea batatas*) is not related to the common potato. Also known as the yam (which it is not) or Spanish potato, this tuberous root is from Central America (and possibly originates from Asia) and was eaten in Europe well before the true potato. (Christopher Columbus brought it to Europe from the Americas in 1493.)

Glazed sweet potatoes are an integral part of the menu at Thanksgiving in the United States and both red and white types are available. The Tudors considered them to be an aphrodisiac. Whether effective in this area or not, there is no doubt that they contain high levels of beta-carotene and are excellent sources of iron, potassium and vitamin C.

GIANT VEGETABLES

When considering the topic of giant vegetables, perhaps the question one should ask is not how big they are, but why grow them in the first place? In countries all around the world there are dedicated groups of gardeners devoting huge amounts of energy to growing onions the size of a man's head, or marrows the height of a teenager. To list current records here would only invite contradiction and revision. As I write, there will be giant marrow growing in all corners of the planet, swelling to assume grotesque and gargantuan forms (indeed, 50kg/110lb is not unknown), or monstrous carrots weighing in at 9kg/20lb plus. What for, I wonder? Certainly not taste. As all home growers know, the most delicious vegetables are those that are picked small and tender, not huge and as tough as the proverbial old boot. I put out a clarion call to all vegetable growers to remember that small is beautiful (and tastes a lot better, too!).

PLANTING POTATOES

When to plant your first potatoes is a vexed question for many gardeners. Old English folklore gives some guidance, although this only works if the potato planter is an adept ornithologist. In Cheshire, the yellow wagtail was known as

the potato dropper, because its first appearance heralded the start of the potato-planting season. Alternatively, others adhered to this rhyme, 'When you hear the cuckoo shout, time to plant your tatties out' (tatties is dialect for potatoes).

THE AUBERGINE
Solanum melongena

SOLANACEAE

THE AUBERGINE is an ancient vegetable, most likely it originated in India, and it has also been cultivated in Africa, China and the Near East for centuries. The vegetable is mentioned in Chinese literature dating from the 5th century and it is thought this is where the smaller fruiting varieties originated. The aubergine appears in northern Europe during the 16th century, although it is quite possible that they may have been grown in hotter areas, such as southern Spain, much earlier than this. As a member of the nightshade family, the first aubergines to be imported were regarded with some suspicion by cautious Europeans as they were thought to be poisonous (just as tomatoes and potatoes were). The Spanish introduced them into America under the name of berengenas, or 'apples of love'. Other names include 'Jew's Apple' and 'Mad Apple' while in India they are known as 'brinjal'. Although the most commonly sold aubergines today are the large, pear-shaped purple ones, home growers can easily find seed of many oval and spherical varieties with green, white, pink, yellow or even striped skins. However, in cooler northern climates they need to be grown in warm conditions under glass.

Solanum macrocarpan and *S. aethiopicum* are the two species of bitter aubergine (sometimes referred to as the 'Tomato of the Jews of Constantine'). Their bitter-tasting fruits are much valued in oriental and African cuisines.

'Purple dwarf aubergine and large purple aubergine', colour lithograph by Faguet (n.d.)

from Album Vilmorin

VILMORIN-ANDRIEUX & CIE (1850–1895)

CALABRESE & BROCCOLI

What is mistakenly often called broccoli (*Brassica oleracea* var. *italica*) is actually calabrese. This has large tight flower heads and quite thick stalks, whereas purple-sprouting broccoli has much smaller and more open flower heads borne on succulent stalks. Both of these brassicas are delicious and, as one might expect with members of the cabbage family, contain high levels of nutrients and are particularly plentiful in cancer-inhibiting elements. To preserve nutrients (and taste) calabrese or broccoli should be lightly steamed and dressed with olive oil and vinegar or lemon juice. Calabrese can also be eaten raw in salads.

Very occasionally one comes across what appears to be miniature sprouting broccoli. This variety goes by a host of names including broccoli rabe, broccoli rape or broccolini.

A BRIEF HISTORY OF LEEKS

The remains of leeks have been uncovered in Egyptian tombs dating back to 1550 BCE, although with their long and fine leaves they look very different from the vegetables we grow today. Likewise, leeks were cultivated by the Romans, and these had slender stems with a very pronounced bulb at the end. It is also thought the Romans ate a hardy leek that still grows wild in parts of the French countryside and is known in Canada and some American states in the north (where it is called 'ramp').

The leek (*Allium porrum*) is a delicious and versatile vegetable, especially if picked young and not overcooked. It is an indispensable ingredient in the famous Scottish soup cock-a-leekie and one of the best heirloom varieties is 'Scotch Flag'. This hails from Musselburgh, Scotland, and was developed by Mr J. Hardcastle in 1834. Consequently it is sometimes known as 'Hardcastle's Musselburgh' or 'Giant Musselburgh'.

VEGETABLE UNIFORMITY

In 2008 the much-ridiculed law passed by the European Union that set 'uniform standardization parameters' for twenty-six types of fruits and vegetables was lifted. The rules had forbidden retailers within the EU from selling such deviants as forked carrots or curved cucumbers, or cauliflowers with a girth of less than 11cm/ 4¹/₃in, and asparagus stalks that were green for less than 80 per cent of their length! Unfortunately, lettuce, tomatoes, sweet peppers along with several fruits remain subject to stringent bureaucratic notions of edible perfection.

HOW TO BLANCH A VEGETABLE

Certain vegetables are blanched while they are growing to improve both their flavour and texture, and the resulting produce is more tender and much less bitter. This is done by protecting the young, tender shoots from light, which results in the plant producing less chlorophyll, thus making them much paler in colour than they would be if grown in the usual way. Among the vegetables that can be treated like this are asparagus, cardoons, celery, chicory, leeks rhubarb and sea kale.

To blanch endives
1. The flavour of young hearts of endives is much improved by simply placing an upturned saucer over the centre of the plant.

To blanch celery
1. Wrap the growing stalks in paper and earth up the soil around the paper.

To blanch leeks
1. There are two ways to blanch leeks. As the young plant develops, gradually cover the white stalk with soil, but only up to where the leaves begin to branch outwards.
2. Alternatively, place a cardboard tube over young leek seedlings when they are about the size of a finger and allow them to grow on inside.

To blanch chicory (the radicchio and sugarloaf varieties)
1. Carefully tie string around the leaves, drawing them together but not too tightly, thereby excluding light from the inner leaves. This must be done approximately ten days prior to harvesting.

To blanch rhubarb
1. In late November pack straw around the plants and cover with a large clay pot or a bucket. This technique keeps out both the light and the winter cold. In these cosy conditions the rhubarb plant is fooled into thinking it is spring and starts to develop brightly coloured shoots, rather than lots of leaves. The resulting slender stalks are especially tender and juicy. Blanching rhubarb is not to be confused with forcing, which involves lifting the root and growing it on in heated, unlit conditions and is usually done on a commercial scale.

THE CELERY

Apium graveolens

A P I A C E A E

WILD CELERY, ancestor to our modern-day stalk celery, was so popular with the Greeks that when they founded Selinunte, Sicily, in 628 BCE, they impressed its image on their coins. The plant was an essential part of the ritual practices of the mythological figure Linus, creator of melody and rhythm. It was cultivated by the Romans and valued for its medicinal properties throughout the Middle Ages. The pale, blanched, stalk variety we recognize as celery today were in cultivation for culinary use in 16th-century Italy and France, then introduced to the New World by the European colonists.

Some modern cultivars are self-blanching but require quite a mild winter; the hardier ones still need to be blanched by mounding up soil around their base. The green types are known as 'Pascal' and the white as 'Golden'. The latter are crisper and less bitter than the green kind. Rather confusingly, a stalk of celery refers to the whole bunch, a single piece is called a rib, and the tender centre ribs are known as the celery heart. These are sometimes sold separately and are considered to be the superior part of the vegetable. It is best to avoid eating the leaves of celery because they often taste bitter and are slightly poisonous. Fresh, crisp celery is best eaten raw, either lightly salted or as crudités with dips. Once celery has begun to wilt, it makes an excellent addition to soups and stews.

'Curl-leaf white celery', colour lithograph by E. Godard (n.d.)

from Album Vilmorin

VILMORIN-ANDRIEUX & CIE (1850–1895)

KNOW YOUR BRASSICAS

Brassica oleracea is just one of the thirty species of the genus *Brassica* and contains many of our most prized (and nutritious) green leafy vegetables. This genus has been divided into seven distinct types as follows:

Brassica oleracea var. *acephala*: curly kale
Brassica oleracea var. *capitata*: red cabbage
Brassica oleracea var. *botrytis*: cauliflower
Brassica oleracea var. *italica*: broccoli
Brassica oleracea var. *gemmifera*: Brussels sprouts
Brassica oleracea var. *gongylodes*: kohlrabi
Brassica oleracea var. *alboglabra*: Chinese broccoli

Among some of the other most popular brassicas is the swede or rutabaga (*Brassica napobrassica*), the turnip (*B. rapa* var. *rapifera*), and bok- or pak-choi (*B. rapa* var. *chinensis*).

BRILLIANT BRUSSELS

Few vegetables provoke such extremes of love and hate as the Brussels sprout (*Brassica oleracea* var. *gemmifera*), and it has been voted Britain's most unpopular vegetable – which may well apply in other countries, too! Scientists have now discovered that what may seem merely a modern-day antipathy might actually date back to our Neanderthal ancestors. They and we apparently share a gene that makes a proportion of the population dislike a bitter chemical called phenylthiocarbamide, which is similar to the plant chemicals that cause the bitter taste in sprouts. This is unfortunate for those who experience this distaste because the nutritional value of sprouts is legendary. Among other beneficial compounds, they contain the vitamins A, B6, C, E, and K, calcium, copper, dietary fibre, folate, iron, manganese, omega-3, and potassium as well as having powerful cancer-preventive qualities.

TO COOK OR NOT

Contrary to much popular opinion, many vegetables actually have a higher nutritional value when cooked than eaten raw. Asparagus, broccoli, cabbage, carrots, peppers, spinach, tomatoes and courgette are among the vegetables that all offer up greater amounts of beneficial antioxidants, such as carotenoids and ferulic acid, if briefly boiled or lightly steamed (though not fried or stir-fried). The downside, however, is that the levels of vitamin C decrease during cooking.

THE BRANDYWINE TOMATO

The heirloom tomato 'Brandywine' is considered by many home growers to be the best cultivar of all, certainly among the large 'beefsteak' type. A potato-leaved plant, it has very large, juicy fruits with a true tomato taste. The vines are vigorous growers and very prolific croppers. The fruits are characterized by their rich red colour, with a slight hint of a purple blush, although an orange-skinned cultivar, confusingly known as 'Yellow Brandywine', is also available. Brandywines have been grown in the United States since at least 1886 and are often listed in seed catalogues as having Amish origins, but this has not been substantiated.

EDIBLE LUFFAS

Something of a vegetable curiosity in Western society, the luffa (*Luffa cylindrica*) is a fast-growing vine plant that needs the warm climate of parts of Asia to thrive. The luffas are only eaten when small and young and look like a cross between a cucumber and a courgette, although they are much shorter and fatter. Once mature, luffas form spongy fibres and after drying, skinning and cleaning they can be used as back scrubs for use in the bathroom or as scouring pads. Other names for these versatile plants are: the dishcloth luffa, sponge gourd and Chinese okra.

RESTORATIVE RADISHES

Today we pick radishes when young and tender, long before they grow too large, turning woody and tough. The Greeks and Romans, though, had other ideas and purportedly harvested radishes that had reached up to 45kg/100lb in weight.

It is possible such giants may have been grown for their alleged medicinal properties, not for tossing in salads. Many claims were made for the curative powers of the radish, including being an effective antidote to poison, removing freckles, curing snake-bites, and even alleviating the pain of childbirth. When mashed up with honey and dried sheep's blood, the radish was also said to restore the hair of those Romans who were bald.

THE LEEK
Allium porrum

ALLIACEAE

TRACES OF THE PRECURSOR of the modern leek have been found in ancient Egyptian tombs and they also formed part of the diet of the Greeks and Romans. Today, the leek is grown for its long white stem that is milder than the onion and lends a subtle flavour to a host of recipes. The leek is the national symbol of Wales and proud Welshmen and women wear it in their hats or on their coats on the 1st of March each year, St David's Day. Legend has it that, as the Welsh went into battle against the Saxons in 640, St David urged them to wear leeks taken from a nearby garden in their hats, thus enabling them to easily distinguish between friend and foe. Because they were victorious, the ploy seems to have worked.

Among the heirloom cultivars that are still available is 'Autumn Giant' (sometimes called 'Autumn Mammoth' or 'Hannibal'). This is very resistant to cold temperatures and has large, fat shafts that are perfect for making hearty winter soups. 'Hilari' has long, firm shafts that are suitable for harvesting in the warmer months when they are still slim and tender.

Many think that all leeks have white shafts topped with dark green strap-like leaves but this is not so. The 19th-century French cultivar 'Large Yellow Poitou' has yellow stems, and the foliage of both 'Blue Green Solaize' and 'Saint Victor' turns violet in cold weather.

'Leek', colour lithograph by Elisa Champin (n.d.)
from Album Vilmorin
VILMORIN-ANDRIEUX & CIE (1850–1895)

THE GLOBE ARTICHOKE
Cynara scolymus

ASTERACEAE

ALTHOUGH CONSIDERED something of an aristocrat of the vegetable world, the beautiful globe artichoke is actually little more than an edible thistle. It was cultivated and consumed as long ago as 2000 BCE in the Middle East and was popular in the kitchens of ancient Rome (when it was believed to be a potent aphrodisiac). The globe artichoke enjoyed a culinary revival during the Renaissance period and features in several 17th-century Italian paintings. It is thought they were introduced into North America by the French settlers in Louisiana and by Spanish adventurers on the west coast.

Artichokes are either green or purple (although both become green once cooked). The purple types are harvested later in the season, are small and tender, and are considered to have the finer taste. Among the purple varietals, 'Violetto' and 'Romanesco' are especially popular, but the aptly named 'Green Globe' is also an established favourite. Baby artichokes can sometimes be found on market stalls – these are the small, immature buds of the plant in which the inner part, the choke, has not yet developed. This hairy choke must be removed in older plants before eating.

Globe artichokes are not to be confused with the Jerusalem artichoke, *Helianthus tuberosus*, a dark-skinned tuber that is not nearly as attractive as its flowering namesake, nor with the tuber Chinese artichoke, *Stachys affinis*, which is white and knobbly.

Cynara scolymus

'Artichoke', by Basilius Besler, with the smaller 'Laon Artichoke', by Elisa Champin

from Hortus Eystettensis and Album Vilmorin

GEORGE MACK (1613), AND VILMORIN-ANDRIEUX & CIE (1850–1895)

KNOW YOUR FENNEL

There are two types of fennel, the one grown for its delicious bulbous root (*Foeniculum vulgare* var. *dulce*) and the tall herb (*F. vulgare*), grown for its leaf and seeds. The bulb type is commonly known as Florence fennel, an indication of its popularity in Italy. It has a tendency to go to seed quite quickly, although if sown in spring, the cultivar 'Zefa Fino' is quite resistant to bolting. If sown close together, the bulbs can be harvested when young and small. To preserve the purity of the various cultivars, grow the herb and vegetable types at some considerable distance from one another, because they are easily cross-pollinated by insects.

HISTORY IN BEANS

The names of some beans commemorate significant moments in history. The climbing French bean 'Cherokee Trail of Tears' evokes the historical 1838 march by the displaced Native American Cherokee nation. It is thought they carried seeds of this very bean with them on their journey to a new homeland. Legend has it that the heirloom bean 'Henderson' was found growing by a road in Virginia by a soldier returning home from the Civil War. (Initially, it was sold commercially as 'Wood's Prolific Bush' then renamed in 1887 when it was sold to Peter Henderson of New York.) On a rather more prosaic note, financially stretched homesteaders and market gardeners sowed the broad bean 'Rent Payer' in hopeful anticipation of bumper crops (much like the fabled prolific tomato 'Mortgage Lifter').

SOME SWEDE FACTS

Known as the swede in Britain, this large root vegetable (*Brassica napobrassica*) is called a rutabaga in North America, and is also referred to as the Swedish turnip or the turnip-rooted cabbage. It is related to the turnip but many cooks think it has a superior taste. It can be cooked in a variety of ways. In the American Midwest, it is often mashed then candied, and in Finland it is casseroled with cream and spices. Originating from northern Europe, the swede is round and coarse-skinned, with sweet-tasting orange flesh and its dark green leaves can be harvested as a cut-and-come-again crop. A useful autumn and winter vegetable, the swede stores well if kept in wooden boxes of moist sand. The heirloom variety 'Wilhemsburger' is a particularly good keeper. Their nutritional value lies in their high concentrations of vitamin C and potassium and their low caloric count.

HOW TO PREPARE & EAT A GLOBE ARTICHOKE

Globe artichokes are the antithesis of fast food, both in preparation and consumption. To grow a good crop requires patience (and space, because a mature plant takes up quite a lot of room in the vegetable plot and, in a good year, will produce a yield of only ten or so artichokes). In the kitchen, as in life, the best things usually come with patience, and the sumptuous artichoke is certainly no exception to this rule. Sadly, too many people are put off cooking and eating globe artichokes, thinking the whole process too much work. However, do try them, following the instructions below, because the taste is well worth it. Due to the time and care needed to eat them, the experience also makes for a very companionable meal to share with friends, the slow and relaxed process being conducive to conversation and shared enjoyment.

1. Soak artichokes upside down in a bowl of salted water for about an hour to remove soil and any hidden insects. If the heads are small and tightly closed, this will not be necessary.
2. Pull the stalk away from the globe head, removing all fibres.
3. Carefully trim the tops of the petal-like leaves with scissors to remove the sharp edges.
4. Slice off the top of the head with a sharp knife.
5. Simmer in a pan of salted water for about 30 minutes, ensuring the heads remain submerged in the water throughout (placing a heavy plate on top can help), by which time the central stalk should pull away easily. Place upside down to drain and cool slightly.
6. Pull out the central core of leaves and keep to one side.
7. Now scoop out the furry core and discard.
8. Replace the central core in the middle and serve.
9. To eat, pull off each leaf individually, dip in melted butter or salad vinaigrette, then nibble on the fleshy base. Discard the tough part of the leaf.

THE BRUSSELS SPROUT

Brassica oleracea var. *gemmifera*

B R A S S I C A C E A E

THE EXACT ORIGIN of the Brussels sprout is obscure, although various theories abound. Certainly it is native to Europe and may have developed from the old and hardy cabbage type known as cottager's kale. The Brussels sprout is recorded in market records dating back to the Middle Ages in Belgium and it appears to have been in cultivation in northern Europe during the 16th and 17th centuries. The much-quoted claim, therefore, that 'the Brussels sprout was discovered in Brussels in 1750' seems somewhat spurious. Although Thomas Jefferson was cultivating sprouts in his large kitchen garden at Monticello, Virginia, in 1812, the vegetable did not gain wide popularity in North America until well into the 1920s.

The delicious, nutrition-packed Brussels sprout is really a miniature cabbage. The tight buds grow all along the length of a thick, strong trunk. The leaves that sprout at the top of the plant can also be harvested and cooked like a cabbage. Sometimes the whole stalk is cut and sold with the sprouts still attached. Most sprouts are green in colour, although the cultivar 'Rubine' has leaves that are tinged with red. Always choose sprouts that are tightly closed; if they are loose leaved and yellowing, then they will be past their best. Briefly cook in lightly salted boiling water until just tender, or steam, but never overcook!

'Brussels sprouts', colour lithograph by Elisa Champin (n.d.)
from Album Vilmorin
VILMORIN-ANDRIEUX & CIE (1850–1895)

MESCLUN

As a definition of mesclun the bland phrases 'mixed salad' or 'salad greens' really do not do it anything like sufficient justice. No two bowls are ever the same; each is unique, wholly dependent on season, location and the mood of its maker. Tender fresh leaves, herbs and flowers all contribute to create a delicious mix of bitter and sweet flavours that are as much a visual and olfactory delight as they are a culinary treat. In Hawaii a similar mix of leaves is known as nalo greens.

POPEYE'S SPINACH

The cartoon character Popeye's propensity for eating spinach boosted US sales of the vegetable by one-third in the early years of the 1930s. So grateful was the spinach industry that several statues of the great man were erected in spinach-growing areas, including Alma in Arkansas, the self-styled 'Spinach Capital of the World'. We may only speculate how much of Popeye's strength and general appeal can be attributed to his impressive spinach consumption. However, the dark leaves are known to be rich in iron, iodine, carotene, folic acid and chlorophyll.

OXHEART VEGETABLES

There are a considerable number of tomatoes known as 'oxhearts'. With their big heart-shaped fruits they make a distinctive and appealing addition to any salad bowl. Among the old heirloom varieties are: 'Oxheart Akers', which has large pink fruits; 'Oxheart Berkshire', from Berkshire in Maine county, USA; and 'Oxheart Hungarian', which boasts an excellent flavour and few seeds. Originally from a small village 30km/20 miles outside Budapest, it had reached the United States by 1902. The Asian cultivar 'Oxheart Japanese' produces tomatoes that can exceed 1kg/2lb in weight. Some other vegetable groups also have oxheart cultivars. Several oval-shaped cabbages are known as oxhearts (sometimes also referred to as pointed heads) including the cultivars 'Arrowhead' and 'Early Jersey Wakefield'. The Oxheart carrot, an old French heirloom, is a stumpy cultivar that can be as wide as it is long.

PEPPER KNOW-HOW

Sweet peppers have hardly any calories but plenty of vitamins.
Green and yellow peppers have high levels of vitamin A while
red ones contain more vitamin C. Chilli peppers are rich in
phytochemicals. Do not discard the inner white membranes
inside a pepper because these are a good source of bioflavonoids,
which promote blood capillary strength.

PREPARING VEGETABLES

Many recipes indicate that vegetables should be treated with proper respect and
consideration when being prepared, rather than simply 'peeled and chopped'. Here are a
few terms one may encounter between the pages of a thoughtful cookery book:

Dice – cut slices lengthways about 1cm/½in thick, then cut into cubes.
Julienne – slice thinly, then cut into strips, similar in length to a matchstick. This method
is particularly suitable for carrots, turnips and leeks; the same result is achieved using a
device known as a mandolin slicer.
Allumettes or **Pailles** – cut into the thickness of a matchstick, then cut into even portions
of about 10cm/2in long. Literally meaning 'matchstick' or 'straw', this technique is usually
used to create superfine potatoes.
Chiffonade – roll up a bunch of leaves then slice across into thin slices. Used to cut very
thinly leafy vegetables, such as spinach and lettuce.

COLOURFUL POTATOES

Not all potatoes have brownish-red skins
and white flesh. Here are just a few of the
rainbow cultivars that could be gracing
your kitchen garden. 'All Blue' is a Peruvian
potato with high nutritional value. 'All
Red', or 'Cranberry Red', has attractive red
skin and flesh. 'Purple Cow Horn' is
flushed with purple. 'Rose Fin Apple',
originating in France, is a waxy-textured
potato with an attractive yellow flesh.
'Salad Blue', has dark blue skin and purple-
blue flesh that turns purple when steamed
or boiled. 'Salad Red', is similar to 'Salad
Blue' but with red flesh. 'Shetland Black' is
a Scottish cultivar with blue skin and
white flesh with a blue ring running
through it. 'Yellow Finn', not surprisingly, is
a cultivar that has yellow flesh.

THE CUCUMBER
Cucumis sativus

THE CUCUMBER is believed to have originated around 3,000 years ago in the Himalayan region of India, where the wild ancestor of our modern-day cucumber, *Cucumis sativus* var. *hardwickii*, still grows today. Archaeological remains of the vegetable, dating back 2,000 years, have been found in Poland and Hungary. It was commonly grown around the Mediterranean region of Europe by the ancient Greeks and the Romans. Indeed, Emperor Tiberius, Pliny the Elder and Columella are all known to have had a penchant for the cooling and refreshing cucumber (over 90 per cent of it is water). Its cultivation in Europe fell into something of a decline until the Renaissance when, as with so many vegetables, its seeds arrived in the New World with Christopher Columbus.

There are various types of cucumber, each with its own particular characteristics. Common outdoor cucumbers have white or black spines. Those cultivated in a greenhouse are smooth-skinned and have been developed from the long-fruited varieties; these are often known as English, hothouse, European, or slicing cucumbers. There are also burpless types, which are less likely to 'repeat' once consumed. The more unusual Sikkim types have red or orange skins. Round cucumbers are often referred to as apple or lemon varieties as they change from a light lemon to a deeper yellow colour on maturity. Smaller ones are known as cornichons or gherkins, and their tart taste makes them perfect for pickling.

'Cucumber', by Ernst Benary (1819–1893)
from Album Benary
COLOUR LITHOGRAPH BY G. SEVEREYNS

THE RUNNER BEAN

Phaseolus coccineus

FABACEAE

IT IS THOUGHT that the runner bean was domesticated much later than the common bean, possibly as late as the third century CE in Mexico. It grows at high altitudes, is tolerant of cool temperatures and heavy rainfall, but dislikes tropical conditions. Known also as the runner in Britain and the United States, the French call it haricot d'Espagne, the Costa Ricans the cuba bean, and in Mexico it has several appellations including *ayocote*, *botil* and *patol*. It was first grown as an ornamental when introduced to 17th-century Europe.

The runner bean is a prodigious climber, quickly smothering a frame or arch with its white, red or bicoloured flowers and long green pods. The beans can be dried and contain a rare amino acid, S methylcisteine, and in Mexico the roots are prized for their medicinal properties. If the bean is to be eaten whole, pod and all, pick them when young, slim and tender. As they mature, the skin toughens and a cordlike string develops along the spine of the pod that should be removed before cooking.

The home grower has a host of heirloom cultivars to choose from. Among them is the 'Aztec Half Runner' (also known as 'Potato Bean' and 'Dwarf White Aztec'), which has large white flowers and white beans. Introduced in 1890, it is thought to originate from the ancient Anasazi and Aztec people. The glamorous sounding 'Black Knight' has red flowers and large black beans.

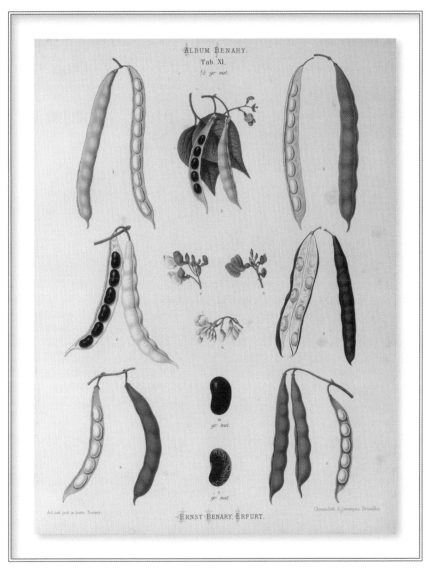

'Runner beans', by Ernst Benary (1819–1893)

from Album Benary

COLOUR LITHOGRAPH BY G. SEVEREYNS

ORNAMENTAL TOMATOES

Tomatoes first arrived in England in the 16th century but it took another three hundred years before cautious Englishmen and women began to eat them with relish (if you'll pardon the pun). Initially, they were grown as ornamental fruits and the yellow, as well as red, cultivars were considered most decorative. Their French name was *pomme d'amour* and it was probably this sobriquet 'love apple' that made the English fearful of their alleged aphrodisiac qualities.

WELSH SAYING

Eat leeks in March and garlic in May,
Then the rest of the year the doctor can play.

SOME CUCUMBER FACTS

In the past, garden legend had it that sowing very fresh seeds produced bitter-tasting and tough-skinned cucumbers. Consequently seeds were often kept for several years. They were then soaked in a variety of preparations, including sheep's milk, honey or mead, before finally being sown. The Romans, apparently, liked their cucumbers straight, without any curve or kink. As soon as the cucumber began to form on the plant, hollow reeds were attached and the cucumber was required to grow inside them, becoming long and regularly shaped. A more sophisticated version of the cucumber trainer was developed by 19th-century gardeners in Britain and was made of glass. Perhaps most oddly of all, the Elizabethans believed that thunder increased the curve of a cucumber!

LUCKY PEAS

According to ancient Norse legend, peas were sent to the Earth by the god Thor; therefore they should only be consumed on his name day, Thursday. When shelling peas look out for pods with only one pea in them, as this is said to be very lucky (though not if you are hungry). Those with nine peas should also bring luck.

NEEPS & TATTIES

'Neep' is the Scots term for the swede, the root vegetable known as rutabaga in the United States (*Brassica napobrassica*). Neeps and tatties (dialect for mashed potatoes) are the traditional accompaniment to haggis, served on Burns Night (January 25). Recipes vary but butter and a little spice such as nutmeg or powdered ginger are common additions. All, of course, must be washed down with a glass of whiskey.

CHANGING COLOURS

Many gardeners love to grow unusually coloured beans, so why only grow the green ones when beautiful yellow and purple cultivars are also available? Yellow and purple pods are also easier to find among the green leaves. However, the colour is often not stable once cooked. 'Purple Queen' and 'Purple Teepee' are purple on the vine but turn green when cooked (although adding a little sugar to the cooking water is said to preserve some of the colour). Likewise, purple sprouting broccoli turns green during the cooking process.

GOOD KING HENRY

Good King Henry (*Chenopodium bonus-henricus*) is a leafy plant that grows wild and is a favourite of foragers, although seeds are available commercially so it can also be grown in the vegetable plot.

The tender young leaves can be eaten raw or cooked, much like spinach, and this highly nutritious plant deserves to be grown more widely. Good King Henry goes by a bewildering range of other common names – among them English mercury, goosefoot, Lincolnshire spinach, spearwort, poor man's asparagus, fat hen, and shoemaker's heels.

Table of Latin & Common Names

The following entries act as a quick reference guide to the scientific names (family, genus, species) and some of the common names for the most popular vegetables.

FAMILY **Alliaceae**

Genus	Species	Common name(s)
ALLIUM	ascalonicum	Shallot
	cepa	Onion Spring onion, green onion, salad onion, scallion
	cepa aggregatum	Potato onion, multiplier onion
	cepa proliferum	Egyptian tree onion, walking onion
	cepa fistulosum	Welsh onion, Japanese onion
	porrum	Leek

FAMILY **Amaranthaceae**

CHENOPODIUM	bonus-henricus	Good King Henry, English mercury, fat hen, goosefoot, Lincolnshire spinach, shoemaker's heels, spearwort
SALICORNIA	europaea	Marsh samphire, glasswort

FAMILY **Apiaceae**

APIUM	graveolens	Celery
	graveolens var. rapaceum	Celeriac, celeri rave, celery root, German celery, knob celery
CRITHMUM	maritimum	Rock samphire, sea asparagus, sea pickle
DAUCUS	carota	Carrot
FOENICULUM	vulgare var. dulce	Fennel, Florence fennel
PASTINACA	sativa	Parsnip
PETROSELINUM	crispum	Hamburg parsley, parsley root, soup parsley, turnip rooted parsley

FAMILY Asteraceae

Genus	Species	Common name(s)
CICHORIUM	*intybus*	Asparagus chicory, pine cone chicory, Catalogna, radichetta
	intybus var. *latifolia*	Chicory, endive, escarole, radicchio, Witloof
CYNARA	*cardunculus*	Cardoon, Texas celery
	scolymus	Globe artichoke
HELIANTHUS	*tuberosus*	Jerusalem artichoke, sunchoke, girasole
LACTUCA	*sativa*	Lettuce
	sativa var. *asparagina*	Asparagus lettuce, celtuce, Chinese lettuce, stem lettuce
SCORZONERA	*hispanica*	Scorzonera, black oyster plant, black salsify, Spanish salsify, viper's grass
TRAGOPOGON	*porrifolius*	Salsify, oyster plant, vegetable oyster

FAMILY Brassicaceae

Genus	Species	Common name(s)
ARUCA	*versicaria*	Arugula, rocket, rocula, rugula
ARMORACIA	*rusticana*	Horseradish
BARBAREA	*praecox*	American cress, Belle Isle, garden, land, winter cress
BRASSICA	*napa* var. *napobrassica*	Swede, rutbaga, Swedish turnip, turnip-rooted cabbage
	oleracea var. *botrytis*	Cauliflower, cabbage flower, coleflower
	oleracea var. *capitata*	Cabbage
	oleracea var. *gemmifera*	Brussels sprout
	oleracea var. *gongylodes*	Kohlrabi, cabbage turnip
	oleracea var. *italica*	Broccoli, calabrese
	rapa var. *rapifera*	Turnip
CRAMBE	*maritima*	Sea kale, chou marin
LEPIDIUM	*sativum*	Cress, English cress, garden cress, land cress
NASTURTIUM	*officinale*	Watercress, brown cress, tall nasturtium
RAPHANUS	*sativus*	Radish
SINAPSIS	*alba*	Mustard

FAMILY Chenopodiaceae

Genus	Species	Common name(s)
ATRIPLEX	*hortensis*	Orache, cabbage of love, good lady of the garden
BETA	*vulgaris*	Beet, Romaine bette
	vulgaris subsp. *cicla*	Chard, leaf spinach, silver beet, white spinach, swiss chard
SPINACIA	*oleracea*	Spinach, mallow of the Europeans

FAMILY Convolvulaceae

IPOMOEA	*batatas*	Sweet potato, yam, Spanish potato

FAMILY Cucurbitaceae

CUCUMIS	*sativus*	Cucumber, gherkin, cornichon
CUCURBITA	*pepo*	Courgette, zucchini
	(many spp.)	Pumpkin, squash
LUFFA	*cylindrica*	Luffa, dishcloth luffa, Chinese okra, sponge gourd

FAMILY Dioscoreaceae

DIOSCOREA	*bulbifera*	Air potato, greater yam, water yam, white yam, winged yam

FAMILY Fabaceae

LOTUS	*tetragonolobus*	Asparagus pea, winged bird's-foot trefoil
PHASEOLUS	*coccineus*	Runner bean, ayocote, botil, cuba, haricot d'Espagne, patol
	lunatus	Butter bean, lima bean
	vulgaris	French bean, haricot, flageolet
PISUM	*sativum*	Pea, English, green, snow, shelling, sugarsnap, mangetout

FAMILY **Fabaceae** *continued*

Genus	Species	Common name(s)
VICIA	*faba*	Broad bean, fava
VIGNA	*unguiculata*	Southern pea, blackeye, cowpea, crowder, yard-long beans

FAMILY **Liliaceae**

ASPARAGUS	*officinalis*	Asparagus, coralwort, sparrow-grass

FAMILY **Malvaceae**

ABELMOSCHUS	*esculentus*	Okra, gumbo, lady's fingers

FAMILY **Poaceae**

ZEA	*mays*	Sweetcorn, maize, corn

FAMILY **Polygonaceae**

RHEUM	*rhaponticum*	Rhubarb, pie-plant

FAMILY **Solanaceae**

CAPSICUM	*annuum*	Bell pepper, wax, cayenne, jalapeños
	chinense	Chilli pepper, habañero, Scotch bonnet
LYCOPERSICON	*lycopersicum*	Tomato, love apple
SOLANUM	*melongena*	Aubergine, apples of love, mad apple, *brinjal*
	macrocarpon and *aethiopicum*	Bitter aubergine
	tuberosum	Potato

Glossary

❧ Annual
A plant is sown, germinates, flowers, sets seed and dies within a single year.

❧ Blanch
Light is excluded from a plant's leaves or stems while it is growing, this keeps them tender and prevents edible plants becoming bitter.

❧ Biodynamic
A system of sowing, planting, pruning and harvesting plants in accordance with the phases of the moon.

❧ Bolt
A plant that bolts produces flowers and seed too early.

❧ Brassica
A plant that belongs to the cabbage family.

❧ Bush
Bush varieties of plants are short stemmed and low growing, as opposed to long-stemmed climbers.

❧ Chitting
The process of exposing seed potatoes to the light, thus encouraging them to develop strong shoots, prior to planting in the ground. Also known as sprouting.

❧ Climber
Climbing varieties of plants produce long stems that twine around tall supports.

❧ Companion planting
The growing of plants alongside each other that are thought to be beneficial to each other either by warding off pests, promoting growth or improving taste.

❧ Crop rotation
The growing of vegetable crops in strict rotation over a three- or four-year period to discourage diseases and pests in the soil.

❧ Cross-fertilization
Also known as cross-pollination, refers to the transfer of pollen from one plant to another, either by insects or human intervention.

❧ Cultivar
A cultivated variety of a plant that has distinct characteristics.

❧ Cut-and-come-again crop
Commonly applied to particular types of lettuce, a cut-and-come-again crop can be harvested by cutting to just above soil level. The plant is then left to resprout and grow for repeated harvests.

❧ Genus
Second most detailed classification (after species, *q.v.*) of a plant, determining the botanical group within the 'family' to which it belongs. The plural form is genera.

❧ Hardening off
The acclimatization of a young plant raised indoors to cooler outdoor conditions by gradual exposure to daytime temperatures.

❧ Earthing up
The process of piling up soil around the base stems of a growing plant. It is used either to blanch the plant or to encourage root formation by preventing them rocking in the wind.

❧ Furrow
A shallow linear depression made in prepared soil prior to planting.

❧ Hardy

A plant that can withstand seasonal changes without protection.

❧ Heirloom or heritage

The terms most commonly used to describe surviving old, open-pollinated cultivar varieties of plants. Most pre-date 1950 and are not grown as part of large-scale commercial agricultural production. Many gardeners now save the seed from their crop to preserve such valuable varieties and swap them with like-minded gardeners. To become a seed saver and swapper, visit www.gardenorganic. org.uk and www.seedsavers.org.

❧ Hot bed

A specially prepared bed for promoting the growth of young plants. It is enriched with well-rotted manure to produce heat. Other methods use glass or plastic sheeting to trap heat from the sun.

❧ Hot house

An artificially heated green or glass house for growing tender and exotic plants.

❧ Legume

A plant that belongs to the Fabaceae or Leguminosae family that have seeds in pods, such as peas and beans.

❧ Maincrop

A vegetable that crops for a long period in the middle of the growing season.

❧ Mesclun

A salad mix made from various leaves and herbs.

❧ Mulch

A thick layer of organic material, such as compost, bark or manure, that is laid on the surface of the earth to retain moisture, suppress weeds and enrich the soil.

❊ Open-pollination
The pollination of plants by natural methods, such as the transference of pollen from one plant to another by insects or the wind.

❊ Perennial
A plant that survives for longer than three seasons, often living for many years.

❊ Pot herb
A herb primarily known for its culinary qualities as opposed to medicinal uses.

❊ Scoville Scale
The heat scale used to measure the concentration of the chemical compound capsaicin in chilli peppers.

❊ Set
The term for the immature bulb of an onion or shallot, or a potato tuber, that is planted to grow on.

❊ Species
Term used to classify, in the greatest detail, an individual type of plant or closely related varieties of plants, that belong to a single genus (*q.v.*). The unique characteristics of the original plant always breed true from seed. The abbreviations 'spp.' and 'subsp.' refer to species (plural) and subspecies respectively.

❊ Successional sowing
The method of sowing seed of fast-maturing vegetables at regular intervals throughout the growing season to provide a continuous supply of fresh produce.

❊ Tap root
A plant's main downward-growing root.

❊ Thinning out
The removal of small seedlings to avoid overcrowding and encourage vigorous growth in the remaining plants.

❊ Tilth
A finely raked surface layer of soil that provides the best conditions for sowing seeds. Also referred to as soil texture.

❊ Variety
Term often used in a loose way to describe a particular cultivar of plant, for instance the 'Painted Lady' variety of runner bean. The abbreviation 'var.' in the scientific name of a plant refers to a division of a species, below the rank of subspecies.

Index

Page numbers in **bold type** refer to colour illustrations.

Acknowledgements & Picture Credits

AUTHOR'S ACKNOWLEDGEMENTS

Many thanks to all at Ivy Press for giving me the opportunity to write about a subject I love, and for their help, support and encouragement during the process. Stephanie Evans must be singled out for special praise for her expertise, patience and forbearance in the face of having to unscramble all those Latin names (not my strong point)!

A big thank you goes to David Wheeler for writing such an engaging and elegant foreword.

I could not have written this book without access to the unrivalled collection of volumes concerning all things horticultural at the Royal Horticultural Society's Lindley Library in London. The shelves of this wonderful resource never fail to yield interesting and unexpected literary gems.

Among my own collection of books the one I turned to again and again was *The Seeds of Kokopelli* by Dominique Guillet. The French Association Kokopelli (known formerly as Terre de Semences Organic Seeds) is an admirable non-profit making organization, dedicated to championing biodiversity and safe guarding traditional fruit and vegetable varieties.

PICTURE CREDITS

The publisher would like to thank the following individuals and organizations for their kind permission to reproduce the images in this book. Every effort has been made to acknowledge the pictures, however we apologize if there are any unintentional omissions.

Akg-images: 2, 9, 13, 17, 21, 25, 29, 33, 35, 39, 41, 45, 49, 53, 57, 61, 65, 69, 73, 75, 79, 83, 87, 91, 95, 99, 103, 105, 109, 115, 113.

Istockphoto/N.Staykov: 27.